PE Mechanical

Machine Design and Materials Practice Exam

Michael R. Lindeburg, PE

PPI®
PPI2PASS.COM

Professional Publications, Inc. • Belmont, California

Benefit by Registering This Book with PPI

- Get book updates and corrections.
- Hear the latest exam news.
- Obtain exclusive exam tips and strategies.
- Receive special discounts.

Register your book at **ppi2pass.com/register**.

Report Errors and View Corrections for This Book

PPI is grateful to every reader who notifies us of a possible error. Your feedback allows us to improve the quality and accuracy of our products. You can report errata and view corrections at **ppi2pass.com/errata**.

PE MECHANICAL MACHINE DESIGN AND MATERIALS PRACTICE EXAM

Current release of this edition: 1

Release History

date	edition number	revision number	update
Oct 2017	1	1	New book.

Printed in the United States of America.

PPI
1250 Fifth Avenue, Belmont, CA 94002
(650) 593-9119
ppi2pass.com

ISBN: 978-1-59126-541-2

Library of Congress Control Number: 2017954342

F E D C B A

Table of Contents

Preface and Acknowledgments

By this time, if you are at a stage in your preparation where you are picking up *PE Mechanical Machine Design and Materials Practice Exam* and reading this preface (most people don't), then you probably have already read the prefaces in the *Mechanical Engineering Reference Manual* and *Practice Problems for the Mechanical Engineering PE Exam.* In that case, you will have read all the clever and witty things I wrote in those prefaces, so I won't be able to reuse them here.

This book accurately reflects the NCEES Mechanical—Machine Design and Materials exam specifications and multiple-choice format. The format of the PE exam presents several challenges to examinees. The breadth of the subject matter covered by the exam requires that you have a firm grasp of mechanical engineering fundamentals. These rudiments are generally covered in an undergraduate curriculum. You still need to know how to select a pump, size a shaft, and humidify a room. However, exam problems enable NCEES to focus on, target, and drill down to some very specific knowledge bases. The problems within each area of emphasis require knowledge gained only through experience—a true test of your worthiness of licensure. Often, problems testing proficiency in an area can be worded as simple definition questions. For example, if you don't recognize the application of a jockey pump, you probably aren't ready to design fire protection sprinkler systems.

Needless to say, the number of problems in this practice exam is consistent with the NCEES exam, with 80 total problems. Like the exam, this book exclusively uses U.S. customary system (USCS) units. The solutions in this practice exam are also consistent in nomenclature and style with the *Mechanical Engineering Reference Manual.* Hopefully, you have already made that book part of your exam-day arsenal.

While *PE Mechanical Machine Design and Materials Practice Exam* reflects the NCEES exam specifications, none of the problems in this book are actual exam problems. The problems in this book have come directly from my imagination and the imagination of my colleagues, Keith Elder, PE, and James Kamm, PhD, PE, LEED AP. Material was technically reviewed by Anthony Darmiento, PE. I am grateful for their efforts.

Editing, typesetting, illustrating, and proofreading of this book continues to follow PPI's strict style guidelines for engineering publications. Managed by Grace Wong, director of publishing services, the following top-drawer individuals have my thanks for bringing this book to life: Nancy Peterson, project manager; Cathy Schrott, production services manager; Jenny Lindeburg King, editor-in-chief; Scott Marley, copy editor; Scott Rutherford, proofreader; Darren Meiss, freelance proofreader; Robert Genevro and Ellen Nordman, typesetters; Tom Bergstrom, technical illustrator and cover designer; and Anil Acharya, calculation checker. Without their efforts, this book would read like a bunch of scribbles.

As in all of my publications, I invite your comments. If you disagree with a solution, or if you think there is a better way to do something, please let me know. You can submit errata online at the PPI website at **ppi2pass.com/errata**.

Best wishes in your exam and subsequent career.

Michael R. Lindeburg, PE

Codes Used to Prepare This Book

The documents, codes, and standards that I used to prepare this book were the most current available at the time of publication. In the absence of any other specific need, that was the best strategy.

Engineering practice is often constrained by law or contract to using codes and standards that have already been adopted or approved. However, newer codes and standards might be available. For example, the adoption of building codes by states and municipalities often lags publication of those codes by several years. By the time the 2015 codes are adopted, the 2017 codes have been released. Federal regulations are always published with future implementation dates. Contracts are signed with designs and specifications that were "best practice" at some time in the past. Nevertheless, the standards are referenced by edition, revision, or date. All of the work is governed by unambiguous standards.

All standards produced by ASME, ASHRAE, ANSI, ASTM, and similar organizations are identified by an edition, revision, or date. However, although NCEES lists "codes and standards" in its lists of Mechanical—Machine Design and Materials exam specifications, no editions, revisions, or dates are specified. My conclusion is that the exam is not sensitive to changes in codes, standards, regulations, or announcements in the Federal Register. This is the reason I referred to the most current documents available as I prepared this book.

Introduction

ABOUT THE PE EXAM

PE Mechanical Machine Design and Materials Practice Exam provides the opportunity to practice taking an eight-hour exam similar in content and format to the Principles and Practice of Engineering (PE) Mechanical—Machine Design and Materials exam. The exam is eight-hours divided into a morning session and an afternoon session. In the four-hour morning session, you are asked to solve 40 problems and in the four-hour afternoon session you are also asked to solve 40 problems. All in all, you will be solving 80 problems in eight hours.

All problems are multiple-choice and in U.S. customary system (USCS) units. They include a problem statement with all the required information, followed by four logical choices. Only one of the four options is correct. The problems are completely independent from each other, so an incorrect choice on one problem will not carry over to subsequent problems.

The problems for each session are drawn from either the Principles or Applications knowledge areas as specified by the National Council of Examiners for Engineering and Surveying (NCEES). The exam knowledge areas and the approximate number of problems are as follows.

Principles (40 problems)

- Basic Engineering Practice (9 problems)
- Engineering Science and Mechanics (10 problems)
- Material Properties (8 problems)
- Strength of Materials (10 problems)
- Vibration (3 problems)

Applications (40 problems)

- Mechanical Components (18 problems)
- Joints and Fasteners (12 problems)
- Supportive Knowledge (10 problems)

HOW TO USE THIS BOOK

This book is a practice exam—the main issue is not how you use it, but when you use it. It was not intended to be a diagnostic tool to guide your preparation. Rather, its value is in giving you an opportunity to bring together all of your knowledge and to practice your test-taking skills. The three most important skills are (1) selection of the right subjects to study, (2) organization of your references and other resources, and (3) time management. Take this practice exam within a few weeks of your actual exam. That's the only time that you will be able to focus on test-taking skills without the distraction of rusty recall.

Do not read the questions ahead of time, and do not look at the answers until you've finished. Prepare for the practice exam as you would prepare for the actual exam. Assemble your reference materials. Check with your state's board of engineering registration for any restrictions on what materials you can bring to the exam. (The PPI website has a listing of state boards at **ppi2pass.com/faqs/state-boards**.)

Read the practice exam instructions (which simulate the ones you'll receive from your exam proctor), set a timer for four hours, and answer the first 40 problems. After a one-hour break, turn to the last 40 problems in this book, set the timer, and complete the simulated afternoon session. Then, check your answers.

The problems in this book were written to emphasize the breadth of the Machine Design and Materials mechanical engineering field. Some may seem easy and some hard. If you are unable to answer a problem, you should review that topic area in the *Mechanical Engineering Reference Manual for the PE Exam.*

The problems are generally similar to each other in difficulty, yet a few somewhat easier problems have been included to expose you to less frequently examined topics. On the exam, only your answer sheet will be scored. No credit will be given for calculations written in your exam booklet.

While some topics in this book may not appear on your exam, the concepts and problem style will be useful practice.

Once you've taken the full exam, check your answers. Evaluate your strengths and weaknesses, and select additional texts to supplement your weak areas (e.g., *Practice Problems for the Mechanical Engineering PE Exam*). Check the PPI website for the latest in exam preparation materials at **ppi2pass.com**.

The keys to success on the exam are to know the basics and to practice solving as many problems as possible. This book will assist you with both objectives.

Pre-Test

You can use the following incomplete table to judge your preparedness. You should be able to fill in all of the missing information. (The completed table appears on the back of this page.) If you are ready for the sample exam (and, hence, for the actual exam), you will recognize them all, get most correct, and when you see the answers to the ones you missed, you'll say, "Ahh, yes." If you have to scratch your head with too many of these, then you haven't exposed yourself to enough of the subjects that are on the exam.

description	value or formula	units
acceleration of gravity, g		in/sec^2
gravitational constant, g_c	32.2	
formula for the area of a circle		ft^2
	1545	ft-lbf/lbmol-°R
	+460	°
density of water, approximate		lbm/ft^3
density of air, approximate	0.075	
specific gas constant for air	53.35	
	1.0	Btu/lbm-°R
foot-pounds per second in a horsepower		ft-lbf/hp-sec
pressure, p, in fluid with density, ρ, in lbm/ft^3, at depth, h		lbf/ft^2
cancellation and simplification of the units (A = amps; rad = radians)	A-sec^4/sec^5-rad	
specific heat of air, constant pressure		Btu/lbm-°R
common units of entropy of steam	–	
	$bh^3/12$	cm^4
molecular weight of oxygen gas		lbm/lbmol
what you add to convert $\Delta T_{°F}$ to $\Delta T_{°R}$		°
	Q/A	ft/sec
inside surface area of a hollow cylinder with length L and diameters d_i and d_o		ft^2
the value that NPSHA must be larger than		ft
	849	lbm/ft^3
what you have to multiply density in lbm/ft^3 by to get specific weight in lbf/ft^3		lbf/lbm
power dissipated by a device drawing I amps when connected to a battery of V volts		W
linear coefficient of thermal expansion for steel	6.5×10^{-6}	
formula converting degrees centigrade to degrees Celsius		
the primary SI units constituting a newton of force		N
the difference between psig and psia at sea level		psi
the volume of a mole of an ideal gas		ft^3
	1.71×10^{-9}	Btu/ft^2-hr-°R^4
universal gas constant	8314.47	
	2.31	ft/psi
shear modulus of steel		psi
conversion from rpm to rad/sec		rad-min/rev-sec
Joule's constant	778.17	

Pre-Test Answer Key

description	value or formula	units
acceleration of gravity, g	386	in/sec^2
gravitational constant, g_c	32.2	lbm-ft/lbf-sec^2
formula for the area of a circle	πr^2 or $(\pi/4)d^2$	ft^2
universal gas constant in customary U.S. units	1545	ft-lbf/lbmol-°R
what you add to $T_{°F}$ to obtain $T_{°R}$ (absolute temperature)	+460	°
density of water, approximate	62.4	lbm/ft^3
density of air, approximate	0.075	lbm/ft^3
specific gas constant for air	53.35	ft-lbf/lbm-°R
specific heat of water	1.0	Btu/lbm-°R
foot-pounds per second in a horsepower	550	ft-lbf/hp-sec
pressure, p, in fluid with density, ρ, in lbm/ft^3, at depth, h	$p=\gamma h=\rho g h/g_c$	lbf/ft^2
cancellation and simplification of the units (A = amps; rad = radians)	A-sec^4/sec^5-rad	W (watts)
specific heat of air, constant pressure	0.241	Btu/lbm-°R
common units of entropy of steam	–	Btu/lbm-°R
centroidal moment of inertia of a rectangle	$bh^3/12$	cm^4
molecular weight of oxygen gas	32	lbm/lbmol
what you add to convert $\Delta T_{°F}$ to $\Delta T_{°R}$	0	°
velocity of flow	Q/A	ft/sec
inside surface area of a hollow cylinder with length L and diameters d_i and d_o	$\pi d_i L$	ft^2
the value that NPSHA must be larger than	h_v (vapor head)	ft
density of mercury	849	lbm/ft^3
what you have to multiply density in lbm/ft^3 by to get specific weight in lbf/ft^3	g/g_c (numerically, 32.2/32.2 or 1.0)	lbf/lbm
power dissipated by a device drawing I amps when connected to a battery of V volts	IV	W
linear coefficient of thermal expansion for steel	6.5×10^{-6}	1/°F or 1/°R
formula converting degrees centigrade to degrees Celsius	°centigrade = °Celsius	The centigrade scale is obsolete.
the primary SI units constituting a newton of force	kg·m/s^2	N
the difference between psig and psia at sea level	14.7 psia (atmospheric pressure)	psi
the volume of a mole of an ideal gas	359 or 360	ft^3
Stefan-Boltzmann constant	1.71×10^{-9}	Btu/ft^2-hr-°R^4
universal gas constant	8314.47	J/kmol·K
conversion from psi to height of water	2.31	ft/psi
shear modulus of steel	11.5×10^6	psi
conversion from rpm to rad/sec	$2\pi/60$	rad-min/rev-sec
Joule's constant	778.17	ft-lbf/Btu

Practice Exam Instructions

In accordance with the rules established by your state, you may use textbooks, handbooks, bound reference materials, and any approved battery- or solar-powered, silent calculator to work this examination. However, no blank papers, writing tablets, unbound scratch paper, or loose notes are permitted. Sufficient room for scratch work is provided in the Examination Booklet.

You are not permitted to share or exchange materials with other examinees.

You will have eight hours for the examination: four hours to answer the first 40 questions, a one hour lunch break, and four hours to answer the second 40 questions. Your score will be determined by the number of questions that you answer correctly. There is a total of 80 questions. All 80 questions must be worked correctly in order to receive full credit on the exam. There are no optional questions. Each question is worth 1 point. The maximum possible score for the examination is 80 points.

Partial credit is not available. No credit will be given for methodology, assumptions, or work written in your Examination Booklet.

Record all of your answers on the Answer Sheet. No credit will be given for answers marked in the Examination Booklet. Mark your answers with a no. 2 pencil. Answers marked in pen may not be graded correctly. Marks must be dark and must completely fill the bubbles. Record only one answer per question. If you mark more than one answer, you will not receive credit for the question. If you change an answer, be sure the old bubble is erased completely; incomplete erasures may be misinterpreted as answers.

If you finish early, check your work and make sure that you have followed all instructions. After checking your answers, you may turn in your Examination Booklet and Answer Sheet and leave the examination room. Once you leave, you will not be permitted to return to work or change your answers.

When permission has been given by your proctor, break the seal on the Examination Booklet. Check that all pages are present and legible. If any part of your Examination Booklet is missing, your proctor will issue you a new Booklet.

WAIT FOR PERMISSION TO BEGIN.

Name: ______________________________
Last First Middle Initial

Examinee number: ______________

Examination Booklet number: ______________

Principles and Practice of Engineering Examination

Practice Examination

1. (A) (B) (C) (D)
2. (A) (B) (C) (D)
3. (A) (B) (C) (D)
4. (A) (B) (C) (D)
5. (A) (B) (C) (D)
6. (A) (B) (C) (D)
7. (A) (B) (C) (D)
8. (A) (B) (C) (D)
9. (A) (B) (C) (D)
10. (A) (B) (C) (D)
11. (A) (B) (C) (D)
12. (A) (B) (C) (D)
13. (A) (B) (C) (D)
14. (A) (B) (C) (D)
15. (A) (B) (C) (D)
16. (A) (B) (C) (D)
17. (A) (B) (C) (D)
18. (A) (B) (C) (D)
19. (A) (B) (C) (D)
20. (A) (B) (C) (D)
21. (A) (B) (C) (D)
22. (A) (B) (C) (D)
23. (A) (B) (C) (D)
24. (A) (B) (C) (D)
25. (A) (B) (C) (D)
26. (A) (B) (C) (D)
27. (A) (B) (C) (D)
28. (A) (B) (C) (D)
29. (A) (B) (C) (D)
30. (A) (B) (C) (D)
31. (A) (B) (C) (D)
32. (A) (B) (C) (D)
33. (A) (B) (C) (D)
34. (A) (B) (C) (D)
35. (A) (B) (C) (D)
36. (A) (B) (C) (D)
37. (A) (B) (C) (D)
38. (A) (B) (C) (D)
39. (A) (B) (C) (D)
40. (A) (B) (C) (D)
41. (A) (B) (C) (D)
42. (A) (B) (C) (D)
43. (A) (B) (C) (D)
44. (A) (B) (C) (D)
45. (A) (B) (C) (D)
46. (A) (B) (C) (D)
47. (A) (B) (C) (D)
48. (A) (B) (C) (D)
49. (A) (B) (C) (D)
50. (A) (B) (C) (D)
51. (A) (B) (C) (D)
52. (A) (B) (C) (D)
53. (A) (B) (C) (D)
54. (A) (B) (C) (D)
55. (A) (B) (C) (D)
56. (A) (B) (C) (D)
57. (A) (B) (C) (D)
58. (A) (B) (C) (D)
59. (A) (B) (C) (D)
60. (A) (B) (C) (D)
61. (A) (B) (C) (D)
62. (A) (B) (C) (D)
63. (A) (B) (C) (D)
64. (A) (B) (C) (D)
65. (A) (B) (C) (D)
66. (A) (B) (C) (D)
67. (A) (B) (C) (D)
68. (A) (B) (C) (D)
69. (A) (B) (C) (D)
70. (A) (B) (C) (D)
71. (A) (B) (C) (D)
72. (A) (B) (C) (D)
73. (A) (B) (C) (D)
74. (A) (B) (C) (D)
75. (A) (B) (C) (D)
76. (A) (B) (C) (D)
77. (A) (B) (C) (D)
78. (A) (B) (C) (D)
79. (A) (B) (C) (D)
80. (A) (B) (C) (D)

Practice Exam

1. Based on chemical resistance alone, which of the following pipe materials would provide the best performance?

(A) polyvinyl chloride (PVC)

(B) polystyrene

(C) polypropylene

(D) acrylonitrile butadiene styrene (ABS)

2. A portion of a pin-connected truss is shown. During an earthquake, pin E experiences a vertical load that alternates between 5 kips upward and 10 kips downward.

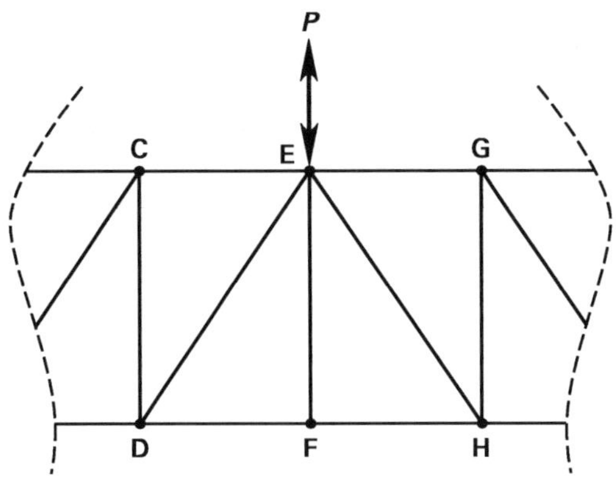

The loading in member EF will

(A) alternate between tension and compression

(B) always be compression

(C) always be tension

(D) always be zero

3. At a remote jobsite, a 12 V direct current inverter provides 120 V alternating current power from a battery to run a 0.6 kW orbital reciprocating saw. The inverter has a resistance of 0.16 Ω and an average efficiency of 68%. What is the approximate effective current flowing to the inverter?

(A) 5.0 A

(B) 7.3 A

(C) 34 A

(D) 74 A

4. The illustration shown is an example of a(n)

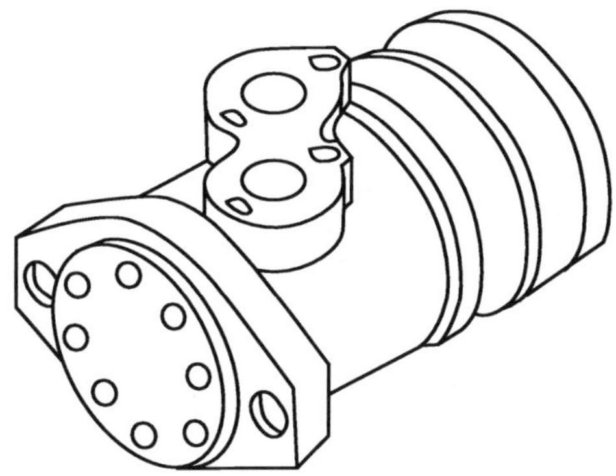

(A) cavalier view

(B) isometric view

(C) principal view

(D) sectional view

5. An aerospace application requires the adhesive bonding of two sheets of titanium skin to form an assembly that can resist relatively high flexural loading. Budgetary restrictions are not a factor. Which of the following types of joints is most appropriate?

(A) double butt lap joint

(B) double scarf lap joint

(C) joggle lap joint

(D) tapered single lap joint

6. The kinematic viscosity of an SAE 10W-30 engine oil with a specific gravity of 0.88 is reported as 110 centistokes (cSt) at 37°C. Expressed in Saybolt universal seconds (SUS), the viscosity is most nearly

(A) 0.90 SUS

(B) 49 SUS

(C) 510 SUS

(D) 32,000 SUS

7. As a project management tool, a graphic record is made of the daily production rate of a construction crew, as shown.

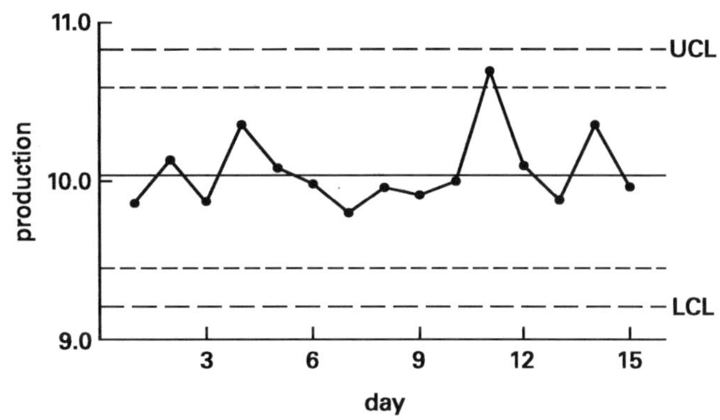

This type of monitoring record is known as a

(A) Gantt chart

(B) moving average chart

(C) *p*-chart

(D) Shewhart chart

8. Investors are evaluating an office building that has a purchase price of $525,000. The investors expect to sell the building for $700,000 after 7 years. During the time that the building is held, the building will be depreciated using straight line depreciation, an estimated life of 15 years, and an (assumed) salvage value of zero. (The full purchase price will be depreciated.) The effective tax rate of the investors is 40%.

During ownership, the estimated expenses will be $25,000 per year, which will be offset by rental income of $45,000 for the first 3 years and $75,000 for the next 4 years.

Neglect the benefit of a favorable capital gains tax rate on the gain above $525,000. The after-tax rate of return on this investment will be most nearly

(A) 2.1%

(B) 7.1%

(C) 7.9%

(D) 12%

9. The end of a circular steel tube is welded to a steel plate. The full-perimeter external fillet weld has a $^3/_{16}$ in throat, and allowable stress at the throat is 10,400 lbf/in^2. The tube outside diameter is 3 in. The tube and plate thicknesses are $^1/_2$ in. A load is applied 4 in from the plate surface, parallel to the plate, perpendicular to the tube axis, and at the centroid of the tube section. Assume the stresses are uniform across the weld. What is most nearly the maximum supported load?

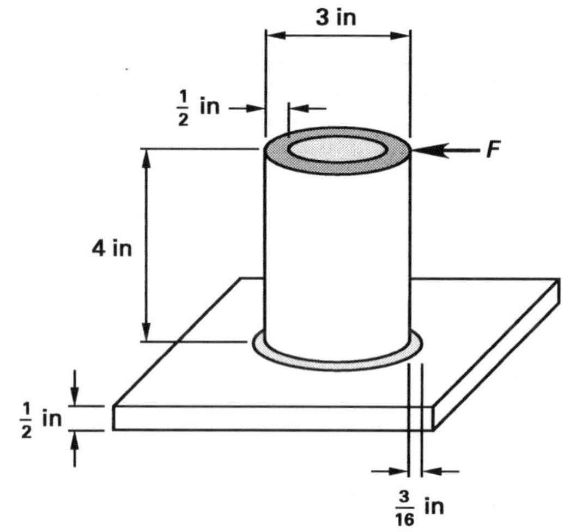

(A) 2900 lbf

(B) 3400 lbf

(C) 5700 lbf

(D) 17,000 lbf

10. A cylindrical tank is produced from alternating layers of a low-ϵ glass, fiber-reinforced epoxy. The fiber orientation angles in alternating layers of the tank wall are at 55° with respect to the principal axes along both the longitudinal and circumferential directions. The tank contains nitrogen gas at a pressure of 10 MPa. The tank has a mean diameter of 300 mm. There is no torsion on the tank. To limit the normal stress to 115 MPa in the direction of fiber orientation, what is most nearly the minimum wall thickness?

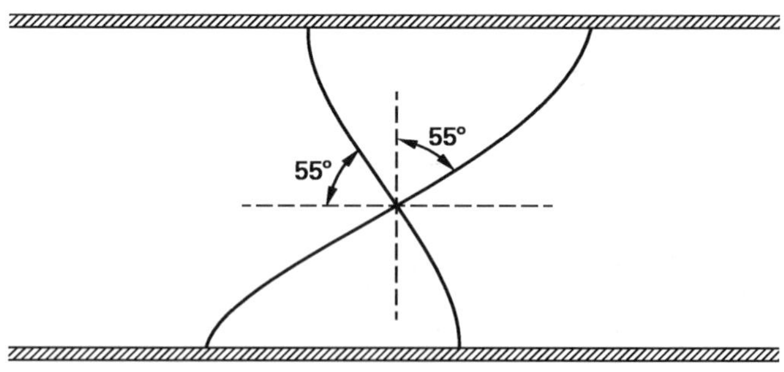

(A) 8.7 mm

(B) 11 mm

(C) 14 mm

(D) 21 mm

11. A welded circular tube with a 0.03 in wall thickness and a 0.50 in outside diameter experiences periodic torque varying between −3 in-lbf and 30 in-lbf. The tube is steel with a 42,000 lbf/in^2 tensile yield strength and a 72,000 lbf/in^2 tensile ultimate strength. The tube endurance strength is 24,000 lbf/in^2, and no stress concentrations exist. What is most nearly the factor of safety for infinite life fatigue loading?

(A) 1.6

(B) 3.2

(C) 4.7

(D) 6.5

12. A 500 g book lays at rest on a broken shelf that has a slope of 37° relative to the bookcase's level base. What is most nearly the minimum coefficient of static friction between the book and the shelf?

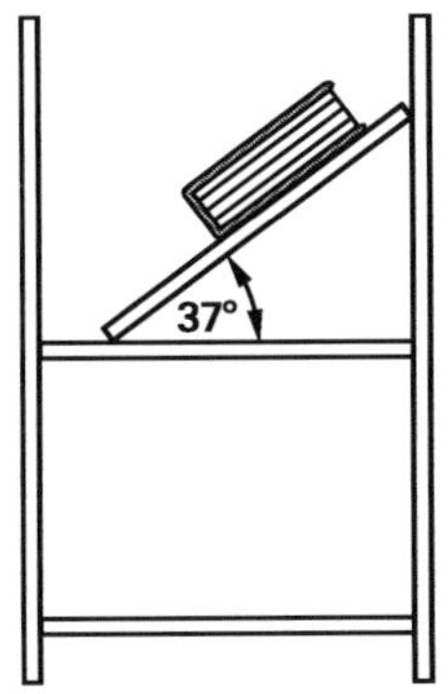

(A) 0.25

(B) 0.37

(C) 0.59

(D) 0.75

13. A cantilevered flat steel spring supports a tip load that periodically varies between 0 lbf and 1 lbf. The spring is 0.75 in wide and 4.0 in long. Tensile yield strength is 75,000 lbf/in^2, tensile ultimate strength is 100,000 lbf/in^2, and endurance strength after accounting for various derating factors is 40,000 lbf/in^2. The design factor of safety is 3. The minimum required spring thickness for infinite life is most nearly

(A) 0.029 in

(B) 0.035 in

(C) 0.041 in

(D) 0.058 in

14. What is the maximum value of dimension A such that the shaft nose will never protrude from the left side of the bearing?

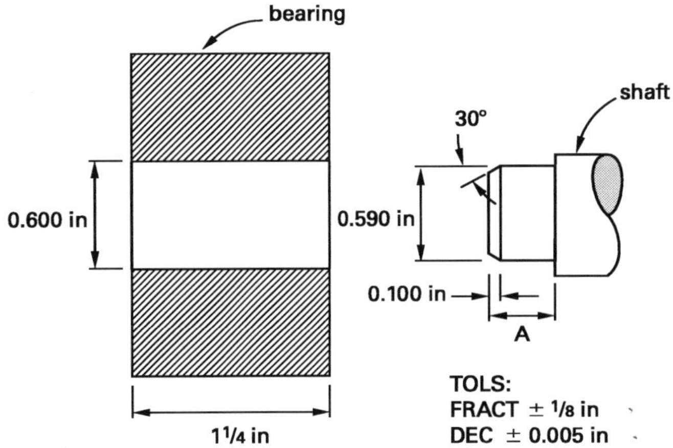

(A) 0.99 in

(B) 1.0 in

(C) 1.1 in

(D) 1.2 in

15. What type of fit is indicated by the manufacturing illustration shown?

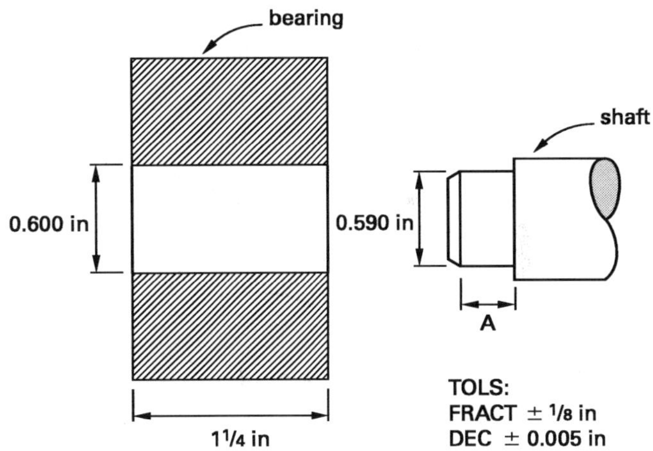

(A) clearance fit

(B) interference fit

(C) press fit

(D) tolerance fit

16. A rotary pendulum comprises a vertical axis, three steel balls spinning around that axis, and an ideal torsional spring resisting the motion. Arms supporting the balls are rigid and massless, the distance from the vertical axis to the ball centers is 0.75 in, and the torsional spring constant is 2.857×10^{-3} in-lbf/rad. The steel ball diameter and density are 1.00 in and 0.284 lbm/in^3, respectively. What is most nearly the natural torsional frequency of the pendulum?

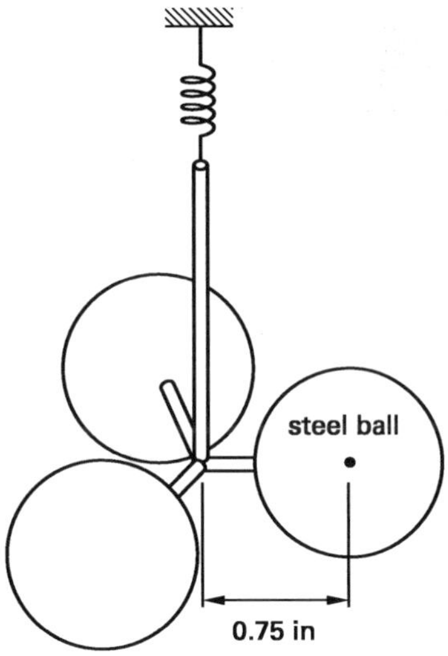

(A) 0.19 Hz

(B) 0.31 Hz

(C) 0.35 Hz

(D) 0.53 Hz

17. A 500 lbm artillery projectile is launched with an initial velocity of 2000 ft/sec, a launch angle of 25° from the horizontal, and an impact elevation of 2500 ft below the launch elevation. Neglecting air resistance, what is most nearly the horizontal travel distance?

(A) 10 mi

(B) 18 mi

(C) 19 mi

(D) 43 mi

18. A titanium tube is loaded in torsion. The 2 ft long tube has a 2 in outside diameter and a 0.125 in wall thickness. The tensile modulus of elasticity is 15.9×10^6 lbf/in^2. The maximum applied torque is 50,000 in-lbf. The maximum angular deflection is most nearly

(A) 0.026 rad

(B) 0.12 rad

(C) 0.31 rad

(D) 0.56 rad

19. A radial ball bearing has a basic load rating of 5660 lbf based on a life of 1,000,000 cycles. The bearing supports a 5000 lbf radial load and a 4000 lbf thrust load. The radial loading factor is 0.56, the thrust loading factor is 1.31, the outer ring rotates, and there is no shock loading. What is most nearly the expected service life?

(A) 280×10^3 cycles

(B) 350×10^3 cycles

(C) 710×10^3 cycles

(D) 840×10^3 cycles

20. A 500 lbf load is eccentrically applied to a 3 in diameter circle of six bolts as shown. What is most nearly the maximum bolt shear load?

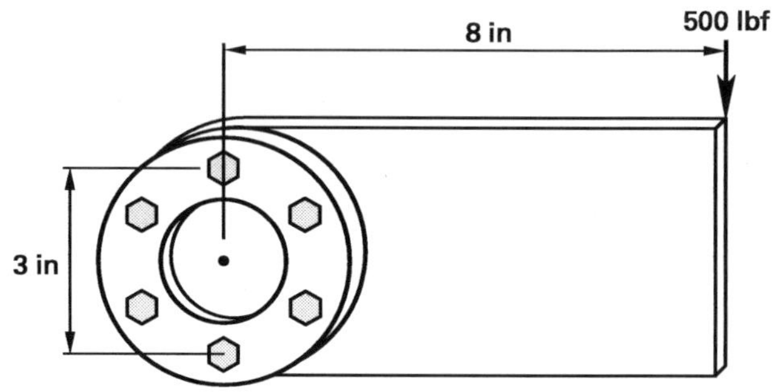

(A) 370 lbf

(B) 450 lbf

(C) 490 lbf

(D) 510 lbf

21. A two-bar linkage of rigid members is rotating clockwise about a fixed point, O, with constant angular velocity of +30 rad/sec as shown. At a particular moment, when member OA forms an angle, θ, of 30° with the horizontal as shown, the horizontal component of the velocity of point B with respect to point A, $v_{B/A,x}$, is −50 in/sec (50 in/sec to the left).

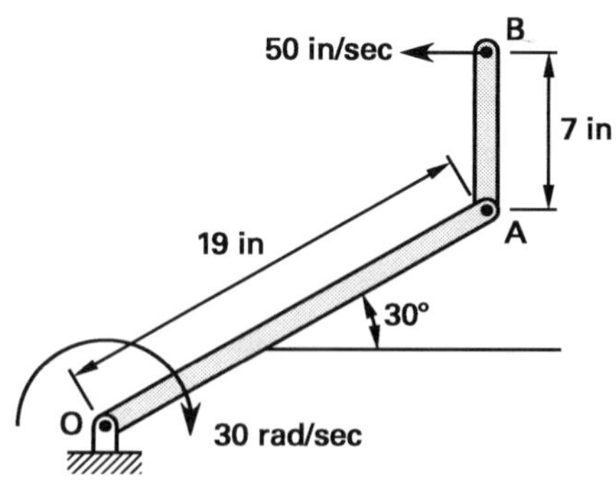

If clockwise rotation is considered positive, the angular velocity of point B with respect to point O, $\omega_{B/O}$, is most nearly

(A) 22 rad/sec

(B) 29 rad/sec

(C) 33 rad/sec

(D) 40 rad/sec

22. A spur gear has a pitch diameter of 4.00 in, a face width of 1.67 in, 34 teeth, and a Lewis form factor of 0.138. Bending stress at the tooth root is limited to 40,000 lbf/in^2. Disregarding stress concentration, what is most nearly the allowable tangential load per tooth?

(A) 3400 lbf

(B) 11,000 lbf

(C) 25,000 lbf

(D) 34,000 lbf

23. A simply supported rectangular-section steel beam is 3.0 in wide, 0.32 in thick, and 35.0 in long. Bending stress is limited to 29,000 lbf/in^2. Deflection is limited to 0.50 in. Neglect buckling. The maximum point load that can be supported at the beam midpoint is most nearly

(A) 65 lbf

(B) 85 lbf

(C) 140 lbf

(D) 170 lbf

24. A helical compression spring made of 0.225 in diameter steel wire has a 2.20 in OD and a 6 in free length. Allowable shear stress is 95,000 lbf/in^2. The spring is dynamically compressed to 150 lbf. What is most nearly the factor of safety?

(A) 0.81

(B) 1.10

(C) 1.20

(D) 1.40

25. A cantilevered beam 40 m long is supported at its left end and 8 m from its right end, as shown. The beam bears a uniform load of 30 N/m and a concentrated load of 400 N applied 6 m from its right end. What is most nearly the vertical shear acting on the beam at point C?

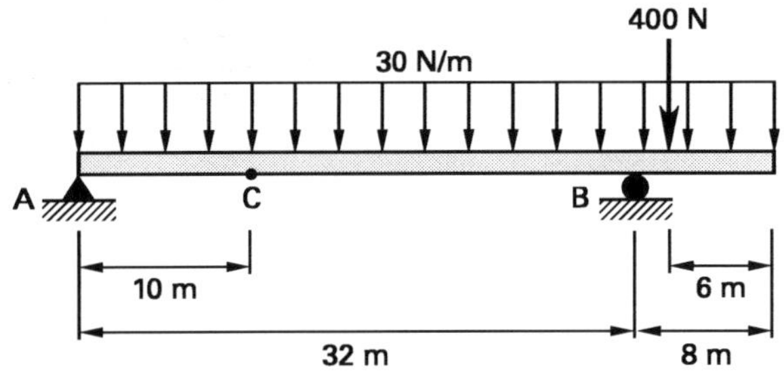

(A) 130 N

(B) 430 N

(C) 750 N

(D) 1200 N

26. A tubular steel shaft (with a modulus of elasticity of 30.5×10^6 psi) has a 2 in OD and a 0.125 in wall thickness. The shaft supports two vertically loaded power transmission components, A and B, as shown. Disregard the mass of the shaft. What is most nearly the shaft's lowest critical speed?

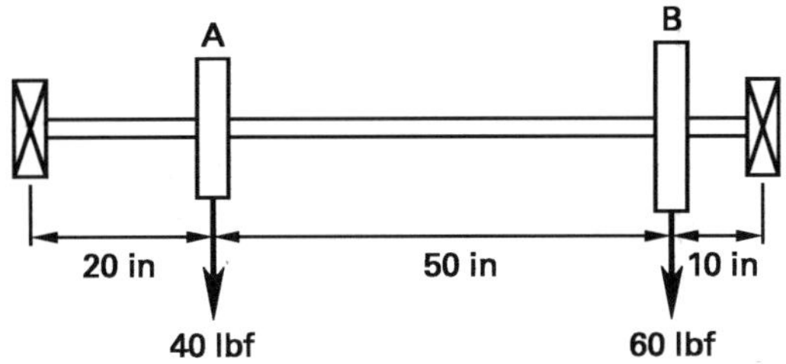

(A) 590 rpm

(B) 940 rpm

(C) 1100 rpm

(D) 1300 rpm

27. A 3.0 in diameter hollow steel shaft is subjected to a static 10,000 in-lbf bending moment and a static 15,000 in-lbf torsional moment. The material yield stress is 80,000 lbf/in^2, and the ultimate stress is 120,000 lbf/in^2. The design factor of safety is 2. The maximum internal diameter is most nearly

(A) 1.9 in

(B) 2.1 in

(C) 2.7 in

(D) 2.9 in

28. The bolted joint shown is designed to carry a shear force of 14,000 lbf and to fail by bolt shear before the force exceeds 20,000 lbf. Characteristics of candidate bolts are listed in the table. There are no threads in the shear plane. Which size bolt best meets the design specifications?

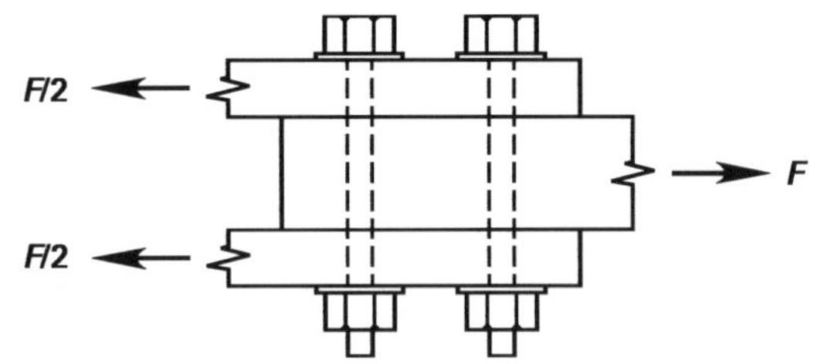

size (in)	major diameter (in)	minor diameter (in)	yield strength (lbf/in^2)	tensile strength (lbf/in^2)
1/4	0.2500	0.1887	57,500	74,500
3/8	0.3750	0.2983	57,500	74,500
1/2	0.5000	0.4056	57,500	74,500
5/8	0.6250	0.5135	57,500	74,500

(A) 1/4 in

(B) 3/8 in

(C) 1/2 in

(D) 5/8 in

29. A nut tightened on a 1″-8UNC-2A bolt develops the full 105,000 lbf/in^2 proof stress. The average radius of nut friction forces is 0.625 in, and all friction coefficients are 0.15. Bolt characteristics are as listed.

pitch diameter	0.9188 in
minor diameter	0.8512 in
lead angle	2.48°
thread half angle	30°
minor diameter area	0.551 in^2
tensile stress area	0.606 in^2

Approximately what amount of torque is required to tighten the nut to full proof stress?

(A) 1000 ft-lbf

(B) 1300 ft-lbf

(C) 1500 ft-lbf

(D) 2400 ft-lbf

30. The four-bar linkage in the illustration represents a mechanism. When the input angle ϕ is 120°, what is most nearly the output angle θ?

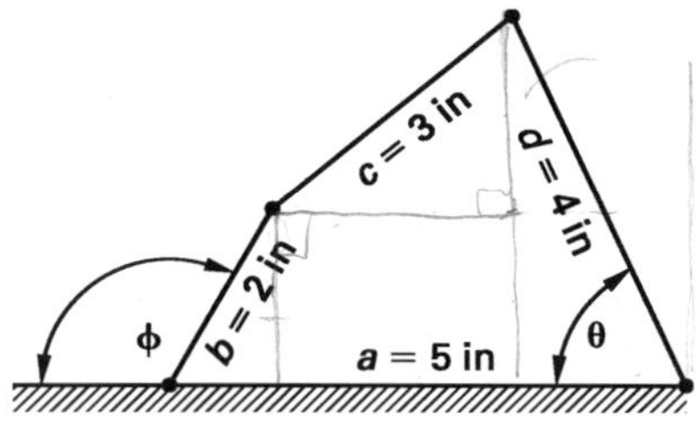

(A) 55°

(B) 65°

(C) 75°

(D) 85°

31. An element of a cast iron component has the static stress state shown. The material is anisotropic with a tensile yield strength of 45,000 lbf/in^2 and a tensile ultimate strength of 65,000 lbf/in^2. The shear yield strength is 10,000 lbf/in^2. Considering only stresses in the x-y plane, what is most nearly the factor of safety?

$$\sigma_x = 10{,}000 \text{ lbf/in}^2$$
$$\sigma_y = 15{,}000 \text{ lbf/in}^2$$
$$\tau_{xy} = \tau_{yx} = 5000 \text{ lbf/in}^2$$

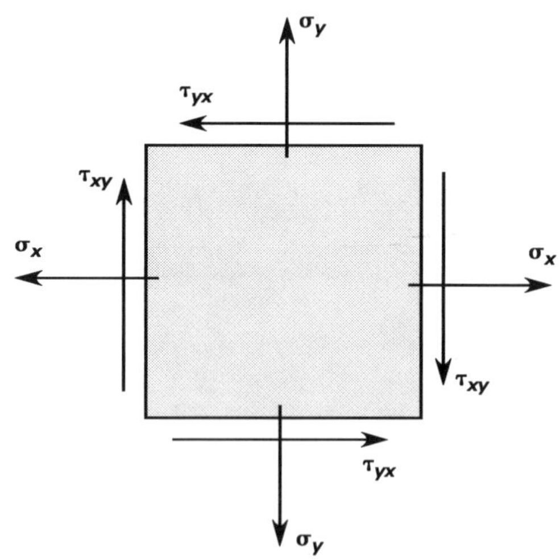

(A) 1.8

(B) 2.5

(C) 2.9

(D) 3.6

32. Design engineers must select a structural polymer from the following list of materials. One critical dimension, fabricated to 10.000 in at 25°C, must not exceed

10.045 in at the operating temperature of 80°C. Material weight must be less than half that of aluminum, which has a specific gravity of 2.71. Which is the best material for the application?

material	yield strength (lbf/in^2)	specific gravity	coefficient of thermal expansion (in/in-°C)	maximum operating temperature (°F)
I	4500	1.05	90×10^{-6}	200
II	10,000	1.40	70×10^{-6}	220
III	8000	1.10	80×10^{-6}	170
IV	9300	1.20	70×10^{-6}	260

(A) I

(B) II

(C) III

(D) IV

33. A machine may be upgraded or replaced. The annualized costs of each option are compared in the table.

	upgrade	replace
annual maintenance	\$500	\$100
initial cost	\$9000	\$40,000
future salvage value	\$10,000 @ year 20	\$15,000 @ year 25
present salvage value	\$13,000	–
interest rate	8%	8%

The uniform annual cost difference between these alternatives is most nearly

(A) \$360

(B) \$750

(C) \$1100

(D) \$2500

34. A 1½ in thick, 12 in OD aluminum disk is shrink-fitted to a 2 in steel shaft with a diametral interference of 0.0045 in. Neglect radial deformation of the shaft during rotation. What is most nearly the rotational speed that will reduce the interference to zero?

(A) 1.6×10^4 rpm

(B) 2.3×10^4 rpm

(C) 1.5×10^5 rpm

(D) 2.2×10^5 rpm

35. Extensive, completely reversed load testing of a nonferrous machine element results in the endurance strength chart shown. Every 9 minutes, the machine element experiences three different loading reversals: $\pm 80{,}000$ lbf/in^2, $\pm 50{,}000$ lbf/in^2, and $\pm 30{,}000$ lbf/in^2. What is most nearly the element fatigue life?

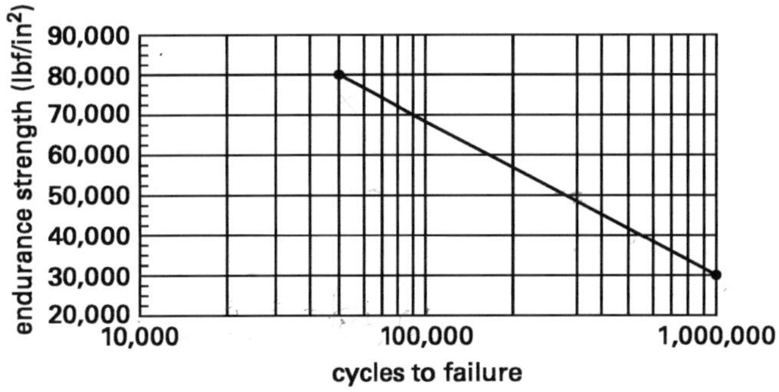

(A) 5500 hr

(B) 6200 hr

(C) 26,000 hr

(D) infinite

36. A 2 in long plastic push rod has a rectangular section measuring 1.0 in × 0.7 in. One end is rigidly fixed in place, while the other end is not supported in any direction. The material's tensile modulus is 420,000 lbf/in^2, and the 2% offset yield strength is 10,000 lbf/in^2. The critical buckling load is most nearly

(A) 5100 lbf

(B) 6100 lbf

(C) 6700 lbf

(D) 7400 lbf

37. A manufacturing process follows a standard normal distribution about the mean. Process results within $\pm 3\sigma$ of the mean are accepted; results beyond these limits are reworked or discarded. The probability that the process will produce an accepted result is most nearly

(A) 68.3%

(B) 95.4%

(C) 99.7%

(D) 99.9%

38. A steel pressure vessel has a 24 in outside diameter, 0.625 in thick shell and elliptical head, 0.5 in thick hemispherical head, and ¹⁄₁₆ in corrosion allowance. Welds are spot radiographed, and operating temperature is 850°F. Allowable stresses versus temperature and weld joint efficiencies are given in the tables. What is most nearly the maximum design pressure?

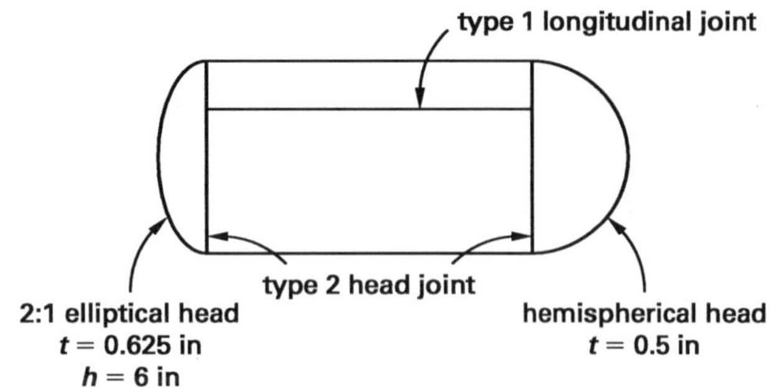

temperature (°F)	allowable stress (lbf/in²)
700	16,700
750	13,900
800	11,400
850	8700
900	5900

structures	weld type	radiography full	radiography spot	radiography none
shells and hemispherical heads	1	1.00	0.85	0.70
shells and hemispherical heads	2	0.90	0.80	0.65
nonhemispherical heads	any	1.00	1.00	0.85

(A) 350 lbf/in²

(B) 450 lbf/in²

(C) 530 lbf/in²

(D) 690 lbf/in²

39. Design requirements specify that a cell phone's plastic shell must survive a 6 ft fall onto concrete. Static compressions tests indicate shell fracture at 1250 lbf and 0.005 in deflection. If a factor of safety of 6 is used, what is most nearly the maximum allowable phone weight?

(A) 0.51 lbf

(B) 0.78 lbf

(C) 0.95 lbf

(D) 1.2 lbf

40. A cam follower for a specific application must have zero acceleration at the start of rise. Which cam profile best meets this requirement?

(A) harmonic

(B) cycloidal

(C) parabolic

(D) velocity derivative

41. A long, thin cantilever beam with a rectangular cross section is subject to lateral vibration. If the thickness of the beam is doubled, the beam's fundamental natural frequency will be

(A) divided by four

(B) divided by two

(C) multiplied by two

(D) multiplied by four

42. European Union directives require many products sold in Europe to display the CE marking. What is this mark?

(A) approval mark issued by a nongovernmental certification body

(B) manufacturer's self-declaration of conformance

(C) government certification signifying successful product testing

(D) mark of origin indicating a product made in the European Community

43. The moment at support A is most nearly

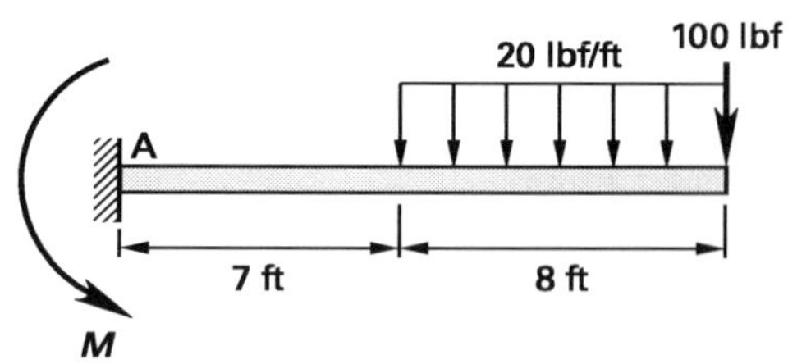

(A) 1900 ft-lbf

(B) 2600 ft-lbf

(C) 3300 ft-lbf

(D) 3900 ft-lbf

44. A 50 lbm mass is connected to the top of a vertical steel rod with a length of 36 in and a diameter of 2 in. The rod is clamped securely at its base, creating an inverted pendulum oscillator. An integral damping system is modeled as a damping coefficient of 1.0 lbf-sec/in. Neglecting the rod's mass, the damping ratio is most nearly

(A) 0.011

(B) 0.036

(C) 0.075

(D) 0.098

45. A tractor pulls a 500 kg crate, initially at rest, up a 26° incline using the cable-pulley system shown. If the tractor moves with a constant velocity of 15 m/s after it is attached to the crate, and the cable remains taut throughout the hauling procedure, what is most nearly the tension in the cable? Neglect the mass of the cable-pulley system, and assume the coefficient of kinetic friction between the crate and ramp is 0.24.

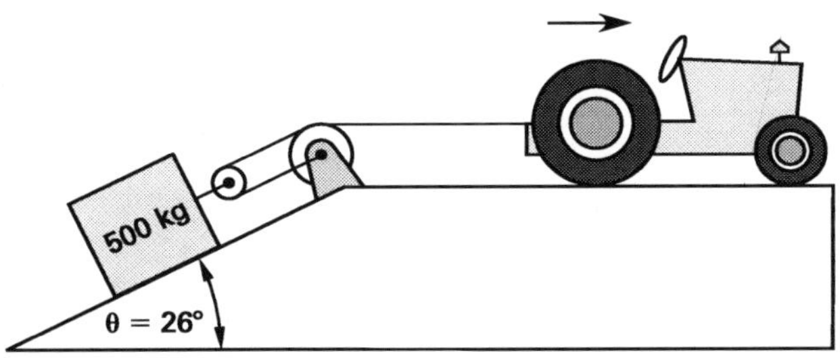

(A) 530 N

(B) 1100 N

(C) 1600 N

(D) 3200 N

46. Zinc blocks attached to the steel hull of a seagoing vessel at various locations below the waterline prevent corrosion. What is the name of this type of corrosion protection?

(A) cathodic protection

(B) anodic protection

(C) passivation

(D) galvanization

47. A machine exerts a force that varies cyclically. At the start of operation, the force increases linearly from zero to 32 lbf over 5 sec. It then decreases linearly from 32 lbf to 20 lbf over 4 sec, and then decreases linearly from 20 lbf to zero in 2 sec, at which time the cycle begins again. The average force during the first 20 sec of operation is most nearly

(A) 13 lbf

(B) 15 lbf

(C) 17 lbf

(D) 19 lbf

48. A solid 90 lbm cylindrical wheel with a radius of 5 ft is rotating at 54 rad/sec. The tangential force that must be applied to the wheel's contact surface in order to reduce the rotational speed by one-third in 30 sec is most nearly

(A) 4 lbf

(B) 10 lbf

(C) 40 lbf

(D) 100 lbf

49. An 18 lbm mass hangs at the end of a wire wrapped around a solid cylinder that is rotating on a frictionless bearing as shown. The cylinder has a mass of 60 lbm and a radius of 15 in.

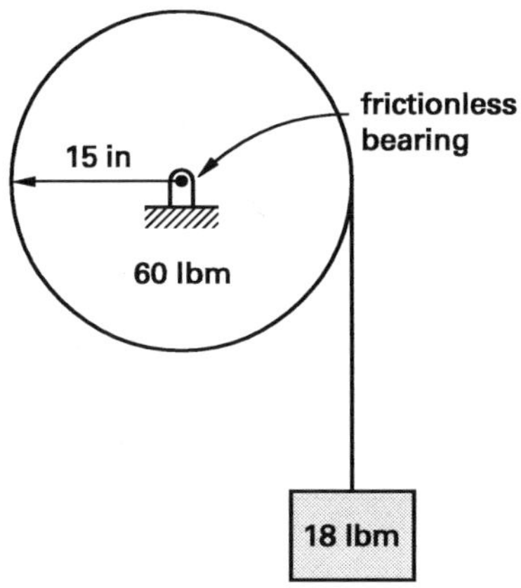

The tension in the wire is most nearly

(A) 11 lbf

(B) 13 lbf

(C) 15 lbf

(D) 17 lbf

50. Aluminum sheet is fillet welded to create the T-section shown. The weld leg size is 1/8 in, sheet thickness is 1/8 in, the weld length is 10 in, and both sides are welded. The allowable force per inch of weld leg is 5000 lbf/in. The joint is required to withstand an impact load of 1000 lbf suddenly applied parallel to the weld and 2 in from the fixed base as shown. What is most nearly the factor of safety?

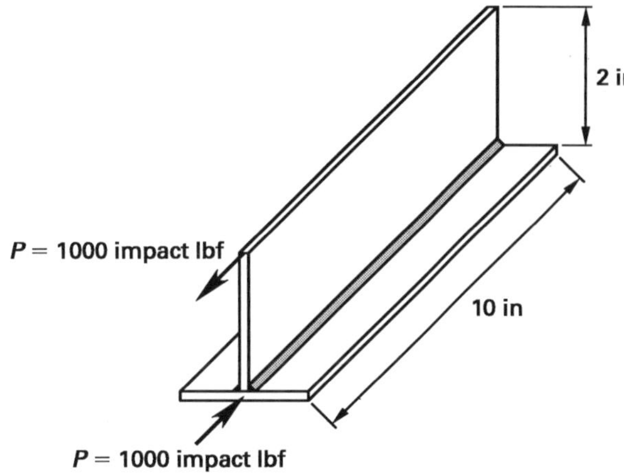

(not to scale)

(A) 4.0

(B) 6.0

(C) 8.0

(D) 16.0

51. A joint is made between two metal pieces by closely fitting their surfaces and distributing a molten nonferrous filler metal to the interface by capillary attraction. The pieces to be joined have a melting point of 1400°F, and the filler metal melts at 700°F. This process is most accurately termed

(A) soldering

(B) brazing

(C) welding

(D) forge welding

52. A 0.500 in nominal diameter hole and pin are to be combined during assembly. Drawings indicate tolerances as listed. The basic hole system is used.

parameter	high tolerance (in)	low tolerance (in)
hole diameter	$\pm 1.6 \times 10^{-3}$ in	-0 in
pin diameter	-2.0×10^{-3} in	-3.0×10^{-3} in

This fit is most properly designated as

(A) RC7 free running fit

(B) LC1 locational clearance fit

(C) LT3 locational-transitional fit

(D) FN2 medium drive fit

53. The motion of a lightly damped system is recorded. The amplitudes of two successive cycles of motion are recorded as 0.569 in and 0.462 in. The damping ratio is most nearly

(A) 0.012

(B) 0.033

(C) 0.068

(D) 0.095

54. A truss is constructed of pin-connected rigid members. The force in member BC is most nearly

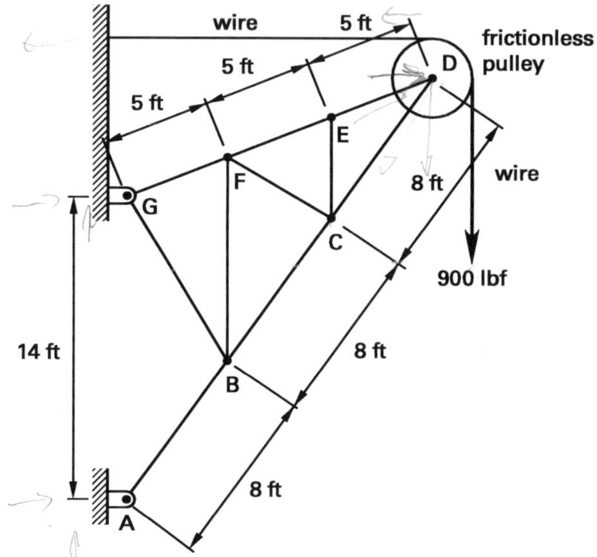

(A) 830 lbf

(B) 880 lbf

(C) 920 lbf

(D) 970 lbf

55. A plant manager is considering replacing an air compressor system with a new 110 hp belt drive model. The compressor will operate at full load for 1800 hr/yr. The new model has an initial cost of \$80,000, an annual maintenance cost of \$2000, and a salvage value of \$10,000 after an expected useful life of 10 yr. The expected interest rate is 10%, and the cost of electricity is \$0.12/kW-hr. Most nearly, the equivalent uniform annual cost for this compressor system is

(A) \$23,000

(B) \$26,000

(C) \$29,000

(D) \$32,000

56. Three parts stacked in assembly have individual nominal dimensions and tolerances as listed. Tolerances are normally distributed and represent $\pm 3\sigma$ from the mean value.

part	dimension (in)
1	0.8 ± 0.1
2	1.0 ± 0.2
3	1.2 ± 0.3

The toleranced height of the stack assembly is most nearly

(A) 3.0 ± 0.06 in

(B) 3.0 ± 0.09 in

(C) 3.0 ± 0.4 in

(D) 3.0 ± 0.6 in

57. Sand castings that have been trimmed and sandblasted are tested for weight before they go on to a machining operation. Every hour a sample is selected, and the average weight of the castings in the sample is recorded. Data for three days of production are shown.

	day 1	day 2	day 3	three-day total	three-day average
hour 1	424	439	421	1284	428.0
hour 2	431	433	440	1304	434.7
hour 3	446	462	444	1352	450.7
hour 4	445	443	444	1332	444.0
hour 5	449	429	452	1330	443.3
hour 6	446	443	446	1335	445.0
hour 7	455	468	452	1375	458.3
hour 8	446	462	450	1358	452.0

Most nearly, the standard deviation of the set of three-day averages is

(A) 4

(B) 6

(C) 8

(D) 10

58. A project consists of seven distinct activities designated as *a* through *g*. The precedence chart for the project is shown.

activity	follows	duration (days)
A		15
B		10
C	A	20
D	B	6
E	B	2
F	C, D	5
G	E, F	8

The length of the critical path is

(A) 24 days

(B) 27 days

(C) 48 days

(D) 52 days

59. Three orthographic views of a yoke plate are shown. All values are in inches.

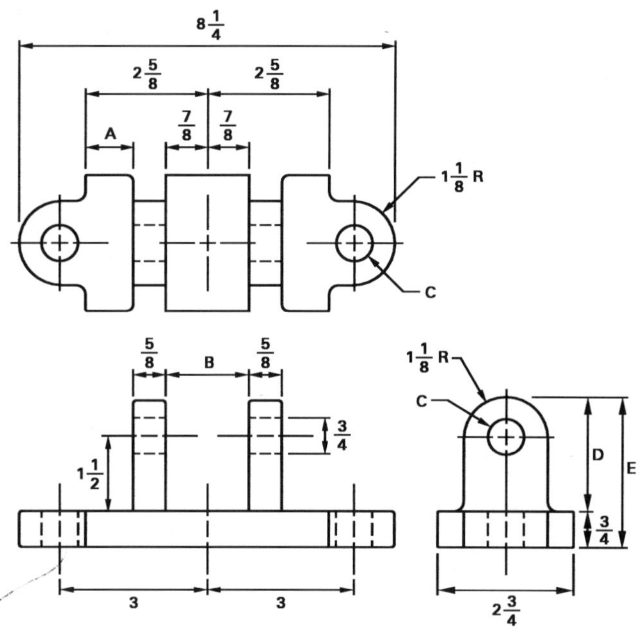

The value of dimension A is

(A) 1 in

(B) 1⅛ in

(C) 1¼ in

(D) 1½ in

60. A section of scaffolding bears loads of 2000 N and 1000 N as shown.

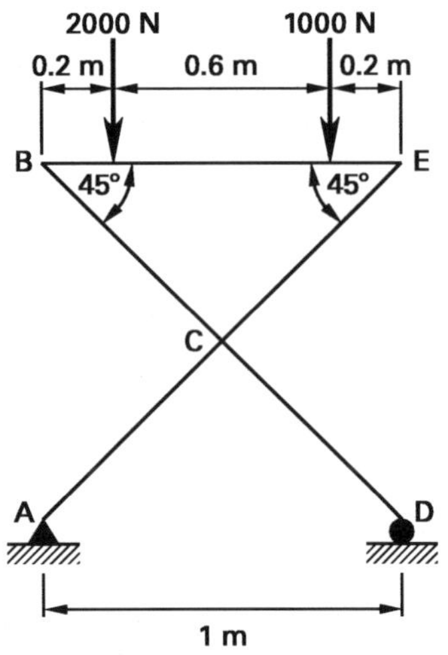

Most nearly, the magnitudes of the horizontal and vertical forces, respectively, on pin E are

(A) 0 N and 1800 N

(B) 1500 N and 1200 N

(C) 1500 N and 1800 N

(D) 3000 N and 1200 N

61. At a shipping facility, two conveyors are used to move boxes. The upper conveyor is horizontal and moves at 9 m/s. At the end of the upper conveyor, the boxes are propelled by inertia and gravity into a curved downward path, and they land on the lower conveyor. The lower conveyor is slanted at a 30° angle from the floor, and the base of the lower conveyor is exactly 10 m below the exit point of the upper conveyor. Air resistance is negligible.

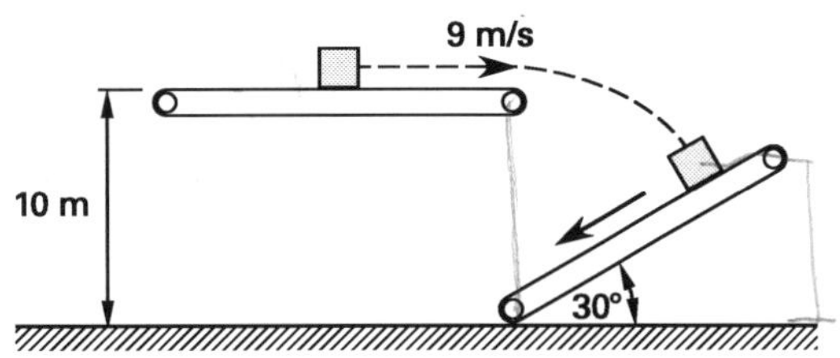

Most nearly, the distance from the base of the lower conveyor to the point where the boxes land on the lower conveyor is

(A) 3.9 m

(B) 5.4 m

(C) 7.5 m

(D) 10 m

62. Which of these metals is a good corrosion-inhibiting alloy for steel?

(A) carbon

(B) manganese

(C) chromium

(D) vanadium

63. Which of these is NOT a scale for classifying the hardness of steel?

(A) Vickers

(B) Brinell

(C) Charpy

(D) Rockwell

64. Poisson's ratio is a measure of a material's

(A) change in density due to temperature change

(B) elongation due to temperature change

(C) elongation due to tension strain

(D) reduction in diameter due to elongation

65. A cylindrical spring with ten coils (all active) is made of steel wire with a diameter of 1 in. The steel has a shear modulus of 12.0×10^6 psi. The design load of the spring is 5000 lbf with a maximum deflection of 6 in. Most nearly, the diameter of the spring is

(A) 3.6 in

(B) 4.6 in

(C) 5.6 in

(D) 6.6 in

66. Two air compression systems are equally capable of performing the desired function, but their costs and efficiencies are different. Each model has an output of 100 hp, and each is expected to have a 10 yr life. Model A has an efficiency of 85%, an initial cost of \$2500, and an annual maintenance cost of \$100. Model B has an efficiency of 92%, an initial cost of \$3300, and an annual maintenance cost of \$60. The expected interest rate is 8%, and the cost of electrical power is \$0.11/kW-hr. Assume that the compressors will always run fully loaded. Most nearly, the number of hours per year that the compressors would have to run to make their costs equivalent is

(A) 110

(B) 140

(C) 170

(D) 220

67. A 120° journal bearing has a diameter of 6 in and a length of 9 in. The bearing's radial clearance is 0.004 in, and its minimum film thickness is 0.002 in. The bearing carries a load of 70 psi of projected bearing area at 1000 rpm. The bearing temperature is 150°F. SAE 10 oil with a specific gravity of 0.85 is used. Most nearly, the frictional torque due to the bearing under these conditions is

(A) 58 in-lbf

(B) 68 in-lbf

(C) 74 in-lbf

(D) 88 in-lbf

68. A 450 lbm flywheel has a diameter of 15 in and a radius of gyration of 6 in. The flywheel is rotating at 500 rpm and is then braked to a full stop. Most nearly, the heat dissipated is

(A) 6 Btu

(B) 14 Btu

(C) 30 Btu

(D) 66 Btu

69. The flywheel in a clutch and flywheel system has a diameter of 8 in and is made from solid cast iron. The flywheel is rotating at 20,000 rpm. Most nearly, the maximum stress on the flywheel is

(A) 1,400,000 psi

(B) 2,000,000 psi

(C) 2,800,000 psi

(D) 3,800,000 psi

70. Most nearly, what is the pitch of a 7/16-20 in thread?

(A) 0.04 in/thread

(B) 0.05 in/thread

(C) 0.2 in/thread

(D) 0.4 in/thread

71. A welded boiler steam drum of SA-515-60 material is fabricated with an inside radius of 19 in. The weld efficiency is 85%. The drum's maximum operating temperature is 500°F with a yield strength of 20,000 psi, and the maximum pressure is 1000 psi. Most nearly, the required wall thickness of the drum is

(A) 0.86 in

(B) 1.2 in

(C) 1.4 in

(D) 1.8 in

72. The hole pattern for a riveted lap joint using two sizes of rivets is shown.

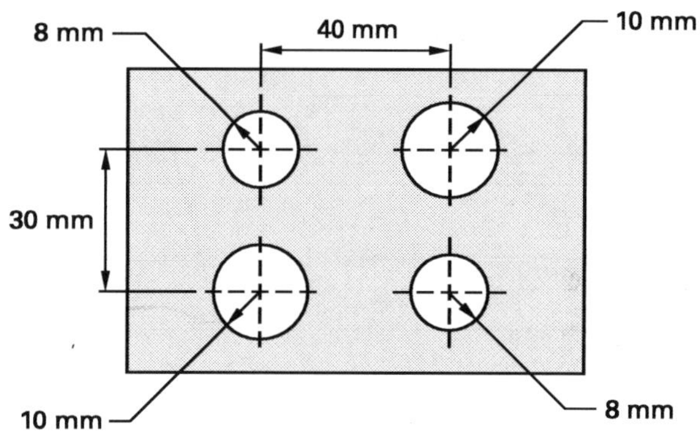

Most nearly, the polar moment of inertia for this pattern around its geometric center is

(A) 0.0016 mm^4

(B) 1.6 mm^4

(C) 160 mm^4

(D) 160,000 mm^4

73. A bolt on a lap joint is preloaded to a tension force of 500 lbf, causing a clamping force on the joint of 500 lbf. The joint is then loaded by a weight of 200 lbf in tension. Most nearly, the clamping force of the joint after the loading is

(A) 0 lbf

(B) 300 lbf

(C) 500 lbf

(D) 700 lbf

74. Two $\frac{1}{4}$ in fillet welds, each 2 in long, carry a steady load of 10,000 lbf parallel to the welds. The weld material has a yield strength of 30,000 psi. Rotational effects are negligible. Most nearly, the factor of safety is

(A) 1.1

(B) 1.4

(C) 1.7

(D) 2.1

75. The cutting tool on a lathe has a cutting speed of 100 ft/min for a rough cut turning operation where the depth of cut is 0.125 in and the feed is 0.02 in/rev. The speed of the lathe (in rpm) is set based on the cutting speed and then never changed. The lathe is used to turn a 1 ft length of round stock with a diameter of 2 in down to a diameter of 1.25 in. Counting only the time the tool is touching the part, and neglecting the time used to set up return passes, the machine time is most nearly

(A) 2.5 min

(B) 4.2 min

(C) 8.6 min

(D) 19 min

76. A milling operation is set up to slot a piece of low-carbon steel using a three-flute, high-speed steel end mill cutter with a diameter of 2 in. The cutting speed for the tool against low-carbon steel is 70 ft/min, and the chip-per-tooth is 0.012 in/tooth. Most nearly, the best feed rate for this operation is

(A) 4.8 in/min

(B) 7.7 in/min

(C) 12 in/min

(D) 20 in/min

77. The following organizations write standards for the machine design industry.

I. American National Standard Institute

II. American Society of Mechanical Engineers

III. American Steel Construction Institute

IV. American Iron and Steel Institute

List, in the following order, the organization that

i. wrote the method of identifying types of steel (e.g., A2360)

ii. wrote the standard for sizing bolts in joints

iii. wrote the standard for pressure vessels

iv. defined fits (e.g., clearance, interference)

(A) I, II, IV, III

(B) II, IV, I, III

(C) III, IV, II, I

(D) IV, III, II, I

78. In the Society of Plastics Industry (SPI) classification system for plastic types, how many code numbers are there?

(A) 7

(B) 11

(C) 16

(D) 24

79. Which of these plastic resins should be marked with this symbol?

(A) acrylonitrile butadiene styrene (ABS)

(B) polystyrene (PS)

(C) polycarbonate (PC)

(D) nylon

80. The x-y home position on a computer numerical control (CNC) mill is $(0, 0)$. However, when measured five times, the x-axis home position was measured as 0.010, 0.014, 0.002, –0.020, and –0.001. Given this data, the best estimate as to how often the home position will differ from zero by more than 0.03 is most nearly

(A) 0.16%

(B) 0.26%

(C) 2.1%

(D) 2.6%

Answer Key

No.	A	B	C	D
1.	A	B	●	D
2.	A	B	C	●
3.	A	B	C	●
4.	A	●	C	D
5.	A	●	C	D
6.	A	B	●	D
7.	A	B	C	●
8.	A	●	C	D
9.	A	●	C	D
10.	A	●	C	D
11.	A	B	C	●
12.	A	B	C	●
13.	A	B	●	D
14.	A	B	●	D
15.	●	B	C	D
16.	A	●	C	D
17.	A	B	●	D
18.	A	B	●	D
19.	●	B	C	D
20.	A	B	C	●
21.	●	B	C	D
22.	●	B	C	D
23.	A	B	●	D
24.	A	B	●	D
25.	●	B	C	D
26.	A	B	●	D
27.	A	B	C	●
28.	A	●	C	D
29.	●	B	C	D
30.	A	●	C	D
31.	●	B	C	D
32.	A	B	C	●
33.	A	B	●	D
34.	●	B	C	D
35.	A	●	C	D
36.	●	B	C	D
37.	A	B	●	D
38.	●	B	C	D
39.	A	B	C	●
40.	A	●	C	D
41.	A	B	●	D
42.	A	●	C	D
43.	A	B	●	D
44.	A	●	C	D
45.	A	B	●	D
46.	●	B	C	D
47.	A	B	C	●
48.	●	B	C	D
49.	●	B	C	D
50.	●	B	C	D
51.	●	B	C	D
52.	●	B	C	D
53.	A	●	C	D
54.	A	B	●	D
55.	A	B	C	●
56.	A	B	●	D
57.	A	B	C	●
58.	A	B	●	D
59.	A	●	C	D
60.	A	B	C	●
61.	A	B	C	●
62.	A	B	●	D
63.	A	B	●	D
64.	A	B	C	●
65.	A	B	●	D
66.	●	B	C	D
67.	A	B	●	D
68.	●	B	C	D
69.	A	B	●	D
70.	A	●	C	D
71.	A	●	C	D
72.	A	B	C	●
73.	A	●	C	D
74.	A	B	C	●
75.	A	B	C	●
76.	●	B	C	D
77.	A	B	C	●
78.	●	B	C	D
79.	A	●	C	D
80.	A	B	●	D

Solutions
Practice Exam

1. PVC, polystyrene, and ABS are examples of amorphous polymers, while polypropylene is a crystalline polymer. In general, crystalline polymers have better chemical resistance than amorphous polymers.

The answer is (C).

2. The truss is pin-connected, and member EF is the only vertical member framing into joint F, so EF is a zero-force member. If member EF were loaded, either in tension or compression, the vertical component of that load could not be transferred out of the joint.

The answer is (D).

3. The reciprocating saw develops 600 W (0.6 kW) of power, which the inverter must supply. The inverter has a 68% efficiency, so the inverter power is

$$P_{\text{inv}} = \frac{P_{\text{saw}}}{\eta} = \frac{600\ \text{W}}{0.68} = 882.4\ \text{W}$$

Method 1

Power equals current times voltage ($P = IV$), so the inverter current is

$$I_{\text{inv}} = \frac{P_{\text{inv}}}{V_{\text{inv}}} = \frac{882.4\ \text{W}}{12\ \text{V}} = 73.5\ \text{A} \quad (74\ \text{A})$$

Method 2

$$\begin{aligned} P_{\text{inv}} &= I_{\text{inv}}V_{\text{inv}} = I_{\text{inv}}(I_{\text{inv}}R_{\text{inv}}) = I_{\text{inv}}^2 R_{\text{inv}} \\ I_{\text{inv}} &= \sqrt{\frac{P_{\text{inv}}}{R_{\text{inv}}}} \\ &= \sqrt{\frac{882.4\ \text{W}}{0.16\ \Omega}} \\ &= 74.3\ \text{A} \quad (74\ \text{A}) \end{aligned}$$

The answer is (D).

4. In a principal, cavalier, or sectional view, at least one projection (also called a projection plane) is parallel with the plane of the screen or paper (or in other words, perpendicular to the direction from which the object is being viewed). The view shown is isometric.

The answer is (B).

5. The types of joints in the four answer options are shown.

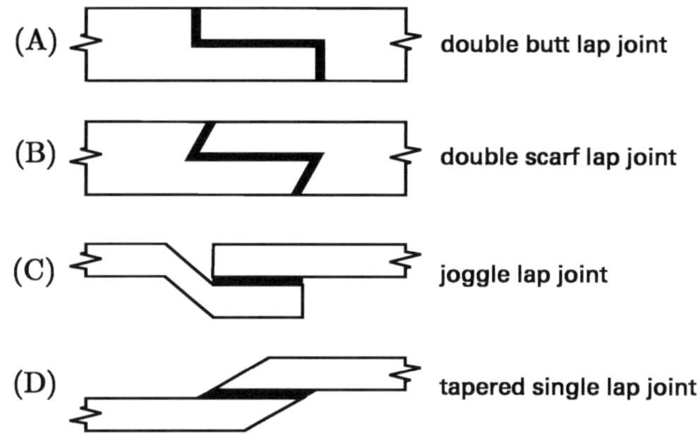

The joggle lap joint (C) can be formed with simple bending operations, without machining. Like a simple lap joint (not shown), the joggle lap joint is relatively subject to peel and cleavage failures. The tapered single lap joint (D), double butt lap joint (A), and double scarf lap joint (B) offer progressively better performance in applications with bending.

The answer is (B).

6. A liquid's kinematic viscosity in Saybolt universal seconds is the time it takes for 60 cm^3 of the liquid to flow through a calibrated tube at a controlled temperature. Intuitively, then, neither 0.90 SUS nor 32,000 SUS is a plausible answer choice for SAE 10W-30 oil.

For viscosities greater than about 240 SUS, one centistoke is equal to about 4.65 SUS. The kinematic viscosity in Saybolt universal seconds is

$$\begin{aligned} \nu &= (110\ \text{cSt})\left(4.65\ \frac{\text{SUS}}{\text{cSt}}\right) \\ &= 512\ \text{SUS} \quad (510\ \text{SUS}) \end{aligned}$$

The answer is (C).

7. The chart shows the day-by-day performance of the production crew. This makes it a *run chart* (also known as a *run sequence chart*). Run charts may or may not have warning and control limit lines, as this one does. A run chart used for quality control purposes is called a *Shewhart chart* (or a *control chart* or a *process-behavior chart*). Run charts were introduced as a quality control tool in the 1920s by Walter Shewhart. The chart can help determine whether day-to-day changes in production are normal variations or indications of an underlying problem that should be identified and corrected.

X-bar (moving average), R (range), and p (proportion) charts are also kinds of Shewhart charts, but the data plotted are different from the data in run charts. In a *moving average chart*, each dot represents average performance over a time period. In a *p-chart*, each dot represents the fraction of the day's production that does not conform to a defined standard. In the chart shown, each dot represents one day's production rate, so this is not a moving average chart or a p-chart.

A *Gantt chart*, which is another project management tool, is a kind of bar chart that shows the start and end dates of elements of a project.

The answer is (D).

8. The interest rate is unknown.

$$D_j = \frac{C - S_n}{n} = \frac{\$525{,}000 - \$0}{15} = \$35{,}000$$
$$\text{BV}_7 = C - jD_j = \$525{,}000 - (7)(\$35{,}000) = \$280{,}000$$

Gain on the sale of a depreciated asset is

$$\text{gain} = S - \text{BV}_7 = \$700{,}000 - \$280{,}000 = \$420{,}000$$

$$\begin{aligned} P = &-\$525{,}000 + (\$700{,}000)(P/F, i\%, 7) \\ &+(\$35{,}000)(P/A, i\%, 7)(0.40) \\ &-(\$420{,}000)(P/F, i\%, 7)(0.40) \\ &-(\$25{,}000)(P/A, i\%, 7)(1 - 0.40) \\ &+(\$45{,}000)(P/A, i\%, 3)(1 - 0.40) \\ &+(\$75{,}000)(P/A, i\%, 7)(1 - 0.40) \\ &-(\$75{,}000)(P/A, i\%, 3)(1 - 0.40) \end{aligned}$$

Neglect the benefit of a favorable capital gains tax rate on the gain above \$525,000.

Simplify and combine terms.

$$\begin{aligned} P &= -\$525{,}000 \\ &\quad + \begin{pmatrix} \$700{,}000 \\ -(\$420{,}000)(0.40) \end{pmatrix} (P/F, i\%, 7) \\ &\quad + \begin{pmatrix} (\$35{,}000)(0.40) \\ -(\$25{,}000)(0.60) \\ +(\$75{,}000)(0.60) \end{pmatrix} (P/A, i\%, 7) \\ &\quad + \begin{pmatrix} (\$45{,}000)(0.60) \\ -(\$75{,}000)(0.60) \end{pmatrix} (P/A, i\%, 3) \\ &= -\$525{,}000 + (\$532{,}000)(P/F, i\%, 7) \\ &\quad +(\$44{,}000)(P/A, i\%, 7) \\ &\quad +(-\$18{,}000)(P/A, i\%, 3) \end{aligned}$$

Try $i = 10\%$.

$$\begin{aligned} P &= -\$525{,}000 + (\$532{,}000)(0.5132) \\ &\quad +(\$44{,}000)(4.8684) \\ &\quad +(-\$18{,}000)(2.4869) \\ &= -\$82{,}532 \end{aligned}$$

Try $i = 5\%$.

$$\begin{aligned} P &= -\$525{,}000 + (\$532{,}000)(0.7107) \\ &\quad +(\$44{,}000)(5.7864) \\ &\quad +(-\$18{,}000)(2.7232) \\ &= \$58{,}676 \end{aligned}$$

Use linear interpolation.

$$\begin{aligned} \text{ROR} &\approx 5\% + (10\% - 5\%)\left(\frac{\$58{,}676}{\$58{,}676 - (-\$82{,}532)}\right) \\ &= 7.08\% \quad (7.1\%) \end{aligned}$$

The answer is (B).

9. Treating the weld as a line, calculate the area, A_w, and section modulus, S_w. (This is equivalent to assuming the weld throat, t_e, is 1.)

$$A_w = \pi D = \pi(3\text{ in}) = 9.425\text{ in}$$

$$\begin{aligned} S_w &= \pi r^2 = \frac{\pi D^2}{4} = \frac{\pi(3\text{ in})^2}{4} \\ &= 7.069\text{ in}^2 \end{aligned}$$

Calculate the shear and bending loads per unit length of weld.

$$\begin{aligned} f_{\text{shear}} &= \frac{V}{A_w} \\ &= \frac{P}{9.425\text{ in}} \\ f_{\text{bending}} &= \frac{M}{S_w} \\ &= \frac{P(4\text{ in})}{7.069\text{ in}^2} \end{aligned}$$

The shear load is assumed to be uniform across the weld. The maximum bending load occurs perpendicular to the applied load, parallel to the tube axis, and at the outside diameter. Combine the shear and bending loads to find maximum load per unit leg.

$$\begin{aligned} f_{max} &= \sqrt{f_{shear}^2 + f_{bending}^2} \\ &= \sqrt{\left(\frac{P}{9.425 \text{ in}}\right)^2 + \left(\frac{P(4 \text{ in})}{7.069 \text{ in}^2}\right)^2} \\ &= \sqrt{(0.106P \text{ in}^{-1})^2 + (0.566P \text{ in}^{-1})^2} \\ &= 0.576P \text{ in}^{-1} \end{aligned}$$

Calculate the allowable load.

$$ft_e = \left(10{,}400 \ \frac{\text{lbf}}{\text{in}^2}\right)\left(\frac{3}{16} \text{ in}\right) = 0.576P \text{ in}^{-1}$$
$$P \le 3385 \text{ lbf} \quad (3400 \text{ lbf})$$

The answer is (B).

10. The hoop stress, or circumferential stress, is

$$\sigma_h = \frac{pr}{t}$$

The longitudinal stress is half this value.

$$\sigma_l = \frac{pr}{2t}$$

The radius of the tank is

$$\begin{aligned} r &= \frac{d}{2} \\ &= \frac{300 \text{ mm}}{2} \\ &= 150 \text{ mm} \end{aligned}$$

The principal stresses are the hoop and longitudinal stresses, and there is no torsion, τ, on the tank, so the stress at an orientation of 55° is

$$\begin{aligned} \sigma_\theta &= \frac{\sigma_h + \sigma_l}{2} + \left(\frac{\sigma_h - \sigma_l}{2}\right)\cos 2\theta + \tau \sin 2\theta \\ &= \frac{\frac{pr}{t} + \frac{pr}{2t}}{2} + \left(\frac{\frac{pr}{t} - \frac{pr}{2t}}{2}\right)\cos 2\theta + \tau \sin 2\theta \\ &= \frac{3pr}{4t} + \left(\frac{pr}{4t}\right)\cos 2\theta + \tau \sin 2\theta \\ \sigma_{55^\circ} &= \frac{3pr}{4t} + \left(\frac{pr}{4t}\right)\cos(2)(55^\circ) + (0 \text{ Pa})\sin(2)(55^\circ) \\ &= 0.665pr/t \end{aligned}$$

The alternating layers are oriented 90° from each other, since the 55° orientation is referenced to both longitudinal and circumferential directions. Check the stress at 90° − 55° = 35°.

$$\sigma_{35^\circ} = \frac{3pr}{4t} + \left(\frac{pr}{4t}\right)\cos(2)(35^\circ) = 0.836pr/t$$

This is a higher stress than in the other direction.

The thickness, t, that is needed to limit the stress, σ_{55°, to 115 MPa is

$$\begin{aligned} t &= \frac{0.836pr}{\sigma_{55^\circ}} \\ &= \frac{(0.836)(10 \text{ MPa})(150 \text{ mm})}{115 \text{ MPa}} \\ &= 10.9 \text{ mm} \quad (11 \text{ mm}) \end{aligned}$$

The answer is (B).

11. The polar moment of inertia is

$$\begin{aligned} J &= \frac{\pi(D^4 - d^4)}{32} = \frac{\pi\left((0.50 \text{ in})^4 - (0.44 \text{ in})^4\right)}{32} \\ &= 0.00246 \text{ in}^4 \end{aligned}$$

The maximum and minimum shear stresses occur at the tube surface.

$$\begin{aligned} \tau_{max} &= \frac{T_{max}r}{J} = \frac{(30 \text{ in-lbf})(0.25 \text{ in})}{0.00246 \text{ in}^4} \\ &= 3049 \text{ lbf/in}^2 \\ \tau_{min} &= \frac{T_{min}r}{J} = \frac{(-3 \text{ in-lbf})(0.25 \text{ in})}{0.00246 \text{ in}^4} \\ &= -304.9 \text{ lbf/in}^2 \end{aligned}$$

The alternating and mean shear stresses are

$$\begin{aligned} \tau_{alt} &= \frac{\tau_{max} - \tau_{min}}{2} = \frac{3049 \ \frac{\text{lbf}}{\text{in}^2} - \left(-304.9 \ \frac{\text{lbf}}{\text{in}^2}\right)}{2} \\ &= 1677 \text{ lbf/in}^2 \\ \tau_m &= \frac{\tau_{max} + \tau_{min}}{2} = \frac{3049 \ \frac{\text{lbf}}{\text{in}^2} + \left(-304.9 \ \frac{\text{lbf}}{\text{in}^2}\right)}{2} \\ &= 1372 \text{ lbf/in}^2 \end{aligned}$$

Calculate the factor of safety using distortion energy theory and the Goodman fatigue line. Using the distortion energy theory, the endurance strength in shear is

$$S_{es} = 0.577S_e = (0.577)\left(24{,}000 \ \frac{\text{lbf}}{\text{in}^2}\right) = 13{,}848 \text{ lbf/in}^2$$

Similarly, the shear ultimate strength is

$$\begin{aligned} S_{us} &= 0.577S_u = (0.577)\left(72{,}000\ \frac{\text{lbf}}{\text{in}^2}\right) \\ &= 41{,}544\ \text{lbf/in}^2 \end{aligned}$$

The factor of safety is

$$\begin{aligned} \text{FS} &= \frac{S_{es}}{\tau_{\text{alt}} + \left(\frac{S_{es}}{S_{us}}\right)\tau_m} \\ &= \frac{13{,}848\ \frac{\text{lbf}}{\text{in}^2}}{1677\ \frac{\text{lbf}}{\text{in}^2} + \left(\frac{13{,}848\ \frac{\text{lbf}}{\text{in}^2}}{41{,}544\ \frac{\text{lbf}}{\text{in}^2}}\right)\left(1372\ \frac{\text{lbf}}{\text{in}^2}\right)} \\ &= 6.49 \quad (6.5) \end{aligned}$$

The answer is (D).

12. For the book-shelf interface, the coefficient of static friction, μ_s, is

$$\mu_s = \tan\theta = \tan 37° = 0.753 \quad (0.75)$$

The answer is (D).

13. The distance to the neutral axis is

$$\begin{aligned} c &= \frac{t}{2} \\ I &= \frac{1}{12}bt^3 = \left(\frac{1}{12}\right)(0.75\ \text{in})t^3 = 0.0625t^3\ \text{in} \end{aligned}$$

The maximum bending stress, σ_{max}, is

$$\begin{aligned} \sigma_{\text{max}} &= \frac{Mc}{I} \\ &= \frac{(1\ \text{lbf})(4.0\ \text{in})\left(\frac{t}{2}\right)}{0.0625t^3\ \text{in}} \\ &= \frac{32\ \text{lbf}}{t^2} \end{aligned}$$

For repeated, one-direction stress,

$$\begin{aligned} \sigma_m = \sigma_{\text{alt}} &= \frac{\sigma_{\text{max}}}{2} = \frac{32\ \text{lbf}}{2t^2} \\ &= \frac{16\ \text{lbf}}{t^2} \end{aligned}$$

Solve using the Goodman equation.

$$\begin{aligned} \text{FS} &= \frac{S_e}{\sigma_{\text{alt}} + \left(\frac{S_e}{S_u}\right)\sigma_m} \\ 3 &= \frac{40{,}000\ \frac{\text{lbf}}{\text{in}^2}}{\frac{16\ \text{lbf}}{t^2} + \left(\frac{40{,}000\ \frac{\text{lbf}}{\text{in}^2}}{100{,}000\ \frac{\text{lbf}}{\text{in}^2}}\right)\left(\frac{16\ \text{lbf}}{t^2}\right)} \\ &= 1786t^2\ \text{in}^{-2} \\ t &= 0.041\ \text{in} \end{aligned}$$

The answer is (C).

14. The width of the bearing is given as 1¼ in, which is a fractional measurement (as opposed to a decimal measurement). The tolerance for fractional measurements is given in the illustration as ⅛ in. In worst-case manufacturing conditions, the bearing would be ⅛ in deficient in width, or 1¼ in − ⅛ in = 1⅛ in, and the shaft nose would be ⅛ in excessive in length.

To keep the shaft nose from protruding even under worst-case conditions, its actual length must be no more than 1⅛ in. The length of the shaft nose is given in decimal measurement, so its tolerance is 0.005 in. In the worst case, the shaft nose would be 0.005 in longer than dimension A, so dimension A can be no more than 1.125 in − 0.005 in = 1.12 in (1.1 in).

The answer is (C).

15. Both the hole in the bearing and the shaft nose are dimensioned in decimals. The tolerance for decimal measurements is given in the illustration as 0.005 in. Under worst-case manufacturing conditions, the bearing hole would be 0.005 in deficient in diameter, or

$$b_{\text{bearing hole,worst}} = 0.600\ \text{in} - 0.005\ \text{in} = 0.595\ \text{in}$$

Under worst-case conditions, the shaft nose would be 0.005 in excessive in diameter, or

$$b_{\text{shaft nose,worst}} = 0.590\ \text{in} + 0.005\ \text{in} = 0.595\ \text{in}$$

The minimum clearance, then, is

$$\begin{aligned} C &= b_{\text{bearing hole,worst}} - b_{\text{shaft nose,worst}} \\ &= 0.595\ \text{in} - 0.595\ \text{in} \\ &= 0\ \text{in} \end{aligned}$$

A fit in which the shaft will always fit inside the hole is called a *clearance fit.* There is no interference between the parts, so this is not an *interference fit* (also known as a *press fit*). The term "tolerance fit" has no specific meaning in engineering.

The answer is (A).

16. Calculate the mass of the three balls.

$$\begin{aligned} m &= 3\rho V = 3\rho \tfrac{4}{3}\pi r^3 \\ &= (3)\left(0.284\ \frac{\text{lbm}}{\text{in}^3}\right)\left(\frac{4}{3}\right)\pi\left(\frac{1.00\ \text{in}}{2}\right)^3 \\ &= 0.446\ \text{lbm} \end{aligned}$$

Calculate the mass moment of a sphere about its centroidal axis.

$$\begin{aligned} I_c &= \tfrac{2}{5}mr^2 \\ &= \left(\frac{2}{5}\right)(0.446\ \text{lbm})\left(\frac{1.00\ \text{in}}{2}\right)^2 \\ &= 0.0446\ \text{lbm-in}^2 \end{aligned}$$

Calculate the rotational moment of inertia using the parallel axis theorem.

$$\begin{aligned} I &= I_c + mx^2 \\ &= 0.0446\ \text{lbm-in}^2 + (0.446\ \text{lbm})(0.75\ \text{in})^2 \\ &= 0.295\ \text{lbm-in}^2 \end{aligned}$$

Calculate the natural frequency.

$$\begin{aligned} \omega &= \frac{1}{2\pi}\sqrt{\frac{kg_c}{I}} \\ &= \frac{1}{2\pi}\sqrt{\frac{\left(2.857\times 10^{-3}\ \frac{\text{in-lbf}}{\text{rad}}\right)\times\left(32.2\ \frac{\text{lbm-ft}}{\text{lbf-sec}^2}\right)\left(12\ \frac{\text{in}}{\text{ft}}\right)}{0.295\ \text{lbm-in}^2}} \\ &= 0.31\ \text{Hz} \end{aligned}$$

The answer is (B).

17. Resolve the initial velocity into horizontal and vertical components.

$$\text{v}_{x,i} = \text{v}_i\cos\theta = \left(2000\ \frac{\text{ft}}{\text{sec}}\right)\cos 25^\circ = 1813\ \text{ft/sec}$$

$$\text{v}_{y,i} = \text{v}_i\sin\theta = \left(2000\ \frac{\text{ft}}{\text{sec}}\right)\sin 25^\circ = 845\ \text{ft/sec}$$

Use the acceleration of gravity and the position equation to find the elapsed time.

$$\begin{aligned} a_{y,i} &= a_{y,f} = g = -32.2\ \text{ft/sec}^2 \\ s_{y,f} &= s_{y,i} + \text{v}_{y,i}t + \tfrac{1}{2}gt^2 \\ -2500\ \text{ft} &= 0\ \text{ft} + \left(845\ \frac{\text{ft}}{\text{sec}}\right)t \\ &\quad + \left(\frac{1}{2}\right)\left(-32.2\ \frac{\text{ft}}{\text{sec}^2}\right)t^2 \end{aligned}$$

Use the quadratic formula.

$$t = 55.3\ \text{sec},\ -2.8\ \text{sec}$$

Using the positive value, calculate the x displacement.

$$\begin{aligned} s_{x,f} &= s_{x,i} + \text{v}_{x,i}t \\ &= \frac{0\ \text{ft} + \left(1813\ \frac{\text{ft}}{\text{sec}}\right)(55.3\ \text{sec})}{5280\ \frac{\text{ft}}{\text{mi}}} \\ &= 19\ \text{mi} \end{aligned}$$

The answer is (C).

18. Find the shear modulus, G, and the polar moment of inertia, J. Poisson's ratio for titanium is approximately 0.34.

$$\begin{aligned} G &= \frac{E}{2(1+\nu)} \\ &= \frac{15.9\times 10^6\ \frac{\text{lbf}}{\text{in}^2}}{(2)(1+0.34)} \\ &= 5.93\times 10^6\ \text{lbf/in}^2 \\ J &= \tfrac{\pi}{32}(D^4 - d^4) \\ &= \left(\frac{\pi}{32}\right)\left((2\ \text{in})^4 - (1.75\ \text{in})^4\right) \\ &= 0.650\ \text{in}^4 \end{aligned}$$

Find the maximum deflection, γ.

$$\begin{aligned} \gamma &= \frac{TL}{JG} = \frac{(50{,}000\ \text{in-lbf})(2\ \text{ft})\left(12\ \frac{\text{in}}{\text{ft}}\right)}{(0.650\ \text{in}^4)\left(5.93\times 10^6\ \frac{\text{lbf}}{\text{in}^2}\right)} \\ &= 0.311\ \text{rad}\quad (0.31\ \text{rad}) \end{aligned}$$

The answer is (C).

19. Compute the equivalent radial load from the larger of the two values. $V_1 = 1.2$, due to outer ring rotation.

$$\begin{aligned} F_e &= V_1 F_r \\ &= (1.2)(5000 \text{ lbf}) \\ &= 6000 \text{ lbf} \\ F_e &= X V_1 F_r + Y F_t \\ &= (0.56)(1.2)(5000 \text{ lbf}) + (1.31)(4000 \text{ lbf}) \\ &= 8600 \text{ lbf} \quad \text{[controls]} \end{aligned}$$

Compute the service life.

$$\frac{L_e}{L} = \left(\frac{C}{F_e}\right)^3$$

$$\begin{aligned} L_e &= L\left(\frac{C}{F_e}\right)^3 \\ &= (1{,}000{,}000 \text{ cycles})\left(\frac{5660 \text{ lbf}}{8600 \text{ lbf}}\right)^3 \\ &= 285 \times 10^3 \text{ cycles} \quad (280 \times 10^3 \text{ cycles}) \end{aligned}$$

The answer is (A).

20. Calculate the reaction due to the applied moment at each bolt.

$$\begin{aligned} M &= Pe \\ &= (500 \text{ lbf})(8 \text{ in}) \\ &= 4000 \text{ in-lbf} \\ R_m &= \frac{M}{Nr} \\ &= \frac{4000 \text{ in-lbf}}{(6 \text{ bolts})(1.5 \text{ in})} \\ &= 444 \text{ lbf/bolt} \end{aligned}$$

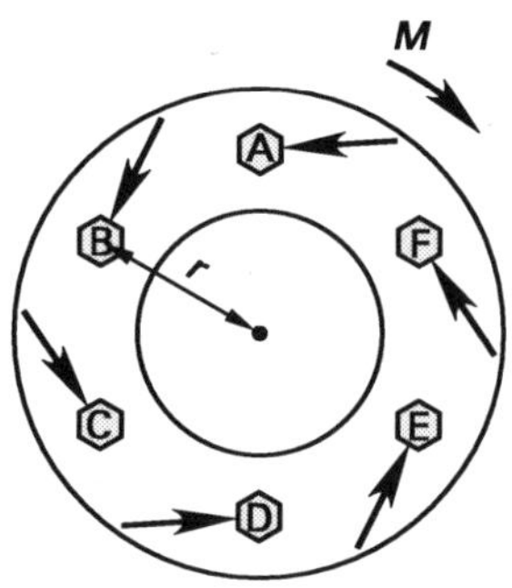

Calculate the reaction due to direct shear at each bolt.

$$\begin{aligned} R_d &= \frac{P}{N} \\ &= \frac{500 \text{ lbf}}{6 \text{ bolts}} \\ &= 83 \text{ lbf/bolt} \end{aligned}$$

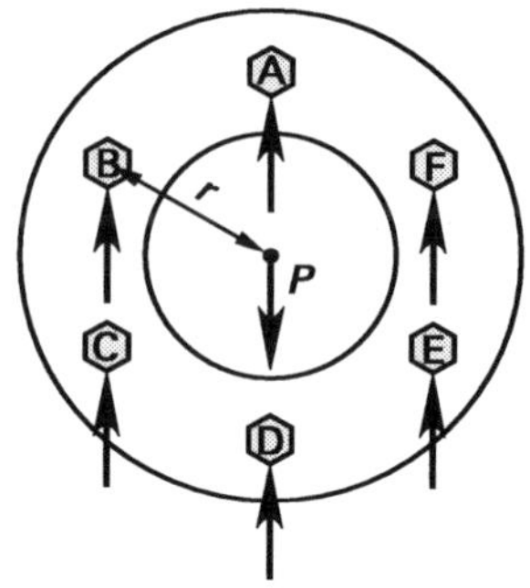

Bolts E and F have the same worst-case reaction combinations. Calculate the vector sum of reactions at bolt E.

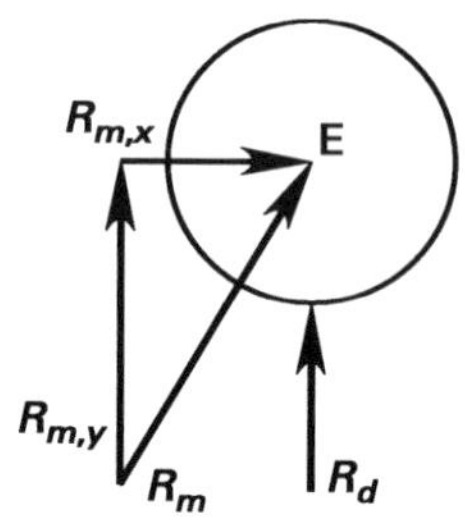

The internal angles are 30°-60°-90°. Calculate the reaction components at bolt E.

$$\begin{aligned} R_x &= R_{m,x} + R_{d,x} \\ &= (444 \text{ lbf})\sin 30° + 0 \text{ lbf} \\ &= 222 \text{ lbf} \\ R_y &= R_{m,y} + R_{d,y} \\ &= (444 \text{ lbf})\cos 30° + 83 \text{ lbf} \\ &= 468 \text{ lbf} \end{aligned}$$

Calculate the maximum total reaction for bolt E.

$$\begin{aligned} R_E &= \sqrt{R_x^2 + R_y^2} \\ &= \sqrt{(222 \text{ lbf})^2 + (468 \text{ lbf})^2} \\ &= 518 \text{ lbf} \quad (510 \text{ lbf}) \end{aligned}$$

The answer is (D).

21. Determine the components of velocity of point A with respect to point O. Use the conventions that x-component velocity is positive to the right and y-component velocity is positive upward.

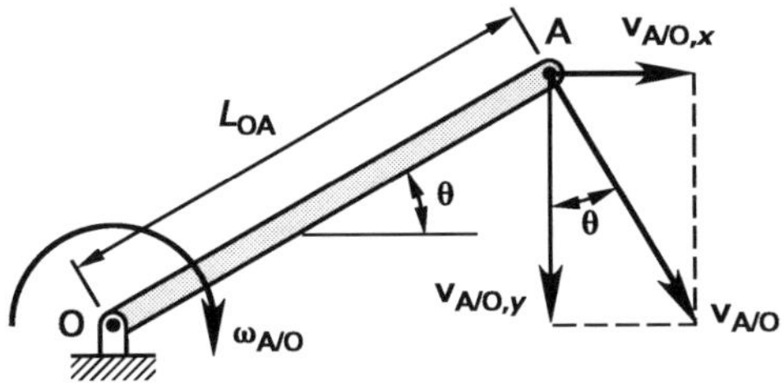

$$
\begin{aligned}
v_{A/O} &= L_{OA}\omega_{A/O} \\
&= (19\text{ in})\left(30\ \frac{\text{rad}}{\text{sec}}\right) \\
&= 570\text{ in/sec} \\
v_{A/O,x} &= v_{A/O}\sin\theta \\
&= \left(570\ \frac{\text{in}}{\text{sec}}\right)\sin 30^\circ \\
&= 285\text{ in/sec} \quad \text{[positive, to the right]} \\
v_{A/O,y} &= -v_{A/O}\cos\theta \\
&= -\left(570\ \frac{\text{in}}{\text{sec}}\right)\cos 30^\circ \\
&= -493.6\text{ in/sec} \quad \text{[negative, downward]}
\end{aligned}
$$

The velocity of point B with respect to point O is the vector sum of the velocity of point A with respect to O and the velocity of point B with respect to A.

$$\mathbf{v}_{B/O} = \mathbf{v}_{A/O} + \mathbf{v}_{B/A}$$

The magnitude of the velocity is calculated from its components. The velocity of point B with respect to point A is negative because it is directed to the left.

$$
\begin{aligned}
v_{B/O,x} &= v_{A/O,x} + v_{B/A,x} \\
&= 285\ \frac{\text{in}}{\text{sec}} + \left(-50\ \frac{\text{in}}{\text{sec}}\right) \\
&= 235\text{ in/sec} \quad \text{[positive, to the right]} \\
v_{B/O,y} &= v_{A/O,y} + v_{B/A,y} \\
&= -493.6\ \frac{\text{in}}{\text{sec}} + \left(-0\ \frac{\text{in}}{\text{sec}}\right) \\
&= -493.6\text{ in/sec} \quad \text{[negative, downward]} \\
v_{B/O} &= \sqrt{v_{B/O,x}^2 + v_{B/O,y}^2} \\
&= \sqrt{\left(235\ \frac{\text{in}}{\text{sec}}\right)^2 + \left(-493.6\ \frac{\text{in}}{\text{sec}}\right)^2} \\
&= 546.7\text{ in/sec}
\end{aligned}
$$

Angle OAB is $30^\circ + 90^\circ = 120^\circ$. From the law of cosines, the length of member OB is

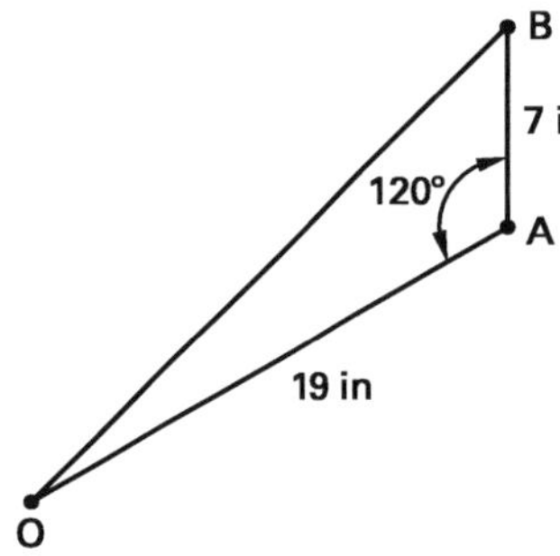

$$
\begin{aligned}
L_{OB} &= \sqrt{L_{OA}^2 + L_{AB}^2 - 2L_{OA}L_{AB}\cos\angle OAB} \\
&= \sqrt{(19\text{ in})^2 + (7\text{ in})^2 - (2)(19\text{ in})(7\text{ in})\cos 120^\circ} \\
&= 23.30\text{ in}
\end{aligned}
$$

Angle OBA is found from the law of sines.

$$
\begin{aligned}
\frac{\sin\angle OBA}{L_{OA}} &= \frac{\sin\angle OAB}{L_{OB}} \\
\angle OBA &= \arcsin\left(\frac{(19\text{ in})\sin 120^\circ}{23.30\text{ in}}\right) \\
&= 44.9^\circ
\end{aligned}
$$

$$
\begin{aligned}
\angle DBE &= \arctan\left(\frac{493.6\ \frac{\text{in}}{\text{sec}}}{235\ \frac{\text{in}}{\text{sec}}}\right) \\
&= 64.54^\circ
\end{aligned}
$$

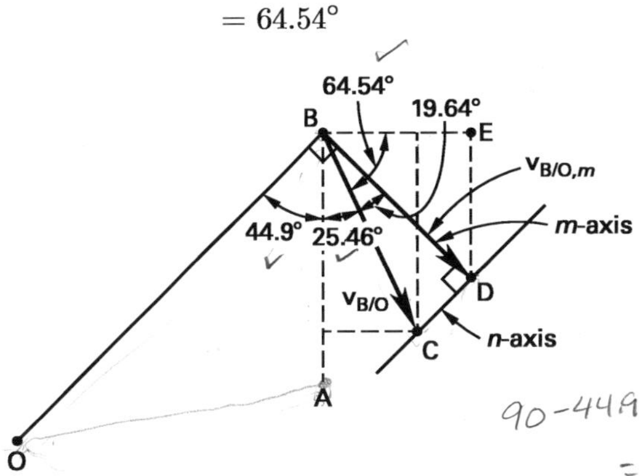

$$
\begin{aligned}
v_{B/O,m} &= v_{B/O}\cos\angle CBD \\
&= \left(546.7\ \frac{\text{in}}{\text{sec}}\right)\cos 19.64^\circ \\
&= 514.9\text{ in/sec}
\end{aligned}
$$

The angular velocity of point B with respect to point O is

$$\omega_{B/O} = \frac{v_{B/O,m}}{L_{OB}} = \frac{514.9\ \frac{\text{in}}{\text{sec}}}{23.20\ \text{in}} = 22.10\ \text{rad/sec} \quad (22\ \text{rad/sec})$$

The answer is (A).

22. Calculate the diametral pitch and form factor.

$$P = \frac{n}{D} = \frac{34\ \text{teeth}}{4.0\ \text{in}} = 8.5\ \text{in}^{-1}$$

$$Y = \pi y = 0.138\pi = 0.434$$

Use the Lewis equation to calculate the allowable tangential load per tooth.

$$F = \frac{\sigma w Y}{P} = \frac{\left(40{,}000\ \frac{\text{lbf}}{\text{in}^2}\right)(1.67\ \text{in})(0.434)}{8.5\ \text{in}^{-1}} = 3411\ \text{lbf} \quad (3400\ \text{lbf})$$

The answer is (A).

23. Find the maximum supported bending load. The distance, c, from the outermost fiber to the neutral plane is

$$c = \frac{h}{2} = \frac{0.32\ \text{in}}{2} = 0.16\ \text{in}$$

The moment of inertia of the beam cross section is

$$I = \frac{bh^3}{12} = \frac{(3.0\ \text{in})(0.32\ \text{in})^3}{12} = 8.19 \times 10^{-3}\ \text{in}^4$$

The maximum moment for a simply supported beam with midpoint load is

$$M_{max} = \frac{PL}{4}$$

The maximum bending load is found by rearranging the bending stress equation.

$$\sigma = \frac{Mc}{I} = \frac{\frac{PL}{4}c}{I}$$

$$P = \frac{4\sigma I}{cL} = \frac{(4)\left(29{,}000\ \frac{\text{lbf}}{\text{in}^2}\right)(8.19 \times 10^{-3}\ \text{in}^4)}{(0.16\ \text{in})(35.0\ \text{in})} = 170\ \text{lbf}$$

The equation for deflection of a simply supported beam with midpoint load is

$$y_{max} = \frac{PL^3}{48EI}$$

The modulus of elasticity for steel is $E = 30 \times 10^6\ \text{lbf/in}^2$. The maximum supported bending load is

$$P = \frac{48EIy_{max}}{L^3} = \frac{(48)\left(30 \times 10^6\ \frac{\text{lbf}}{\text{in}^2}\right)(8.19 \times 10^{-3}\ \text{in}^4)(0.50\ \text{in})}{(35.0\ \text{in})^3} = 137.5\ \text{lbf} \quad (140\ \text{lbf})$$

The answer is (C).

24. Compute the mean diameter, spring index, and Wahl factor.

$$D_m = D_o - D_w = 2.20\ \text{in} - 0.225\ \text{in} = 1.975\ \text{in}$$

$$C = \frac{D_m}{D_w} = \frac{1.975\ \text{in}}{0.225\ \text{in}} = 8.78$$

$$K = \frac{4C-1}{4C-4} + \frac{0.615}{C} = \frac{(4)(8.78)-1}{(4)(8.78)-4} + \frac{0.615}{8.78} = 1.17$$

Compute the maximum stress.

$$\begin{aligned}\tau_{\max} &= \frac{8KFC}{\pi D_w^2} \\ &= \frac{(8)(1.17)(150\text{ lbf})(8.78)}{\pi(0.225\text{ in})^2} \\ &= 77{,}508\text{ lbf/in}^2\end{aligned}$$

Compute the factor of safety.

$$\begin{aligned}\text{FS} &= \frac{\tau_{\text{allowable}}}{\tau_{\max}} \\ &= \frac{95{,}000\ \frac{\text{lbf}}{\text{in}^2}}{77{,}508\ \frac{\text{lbf}}{\text{in}^2}} \\ &= 1.22 \quad (1.20)\end{aligned}$$

The answer is (C).

25. Draw a free-body diagram, including reaction forces at supports A and B.

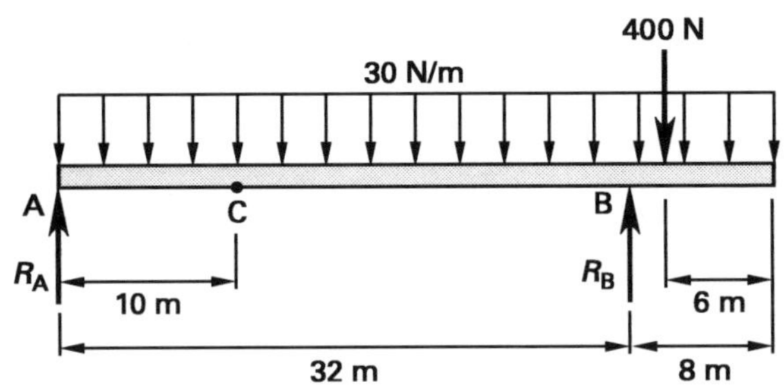

Because the beam is a rigid body at rest, both the sum of the external forces and the sum of the external moments acting on it are zero. The equilibrium equations can be used to solve for the reaction forces.

From the free-body diagram, a moment balance at support A yields

$$\begin{aligned}\sum M_A &= 0 \\ (32\text{ m})R_B - (34\text{ m})(400\text{ N}) \\ -\left(\frac{40\text{ m}}{2}\right)(40\text{ m})\left(30\ \frac{\text{N}}{\text{m}}\right) &= 0 \\ R_B &= 1175\text{ N}\end{aligned}$$

Similarly, at support B,

$$\begin{aligned}\sum M_B &= 0 \\ -(32\text{ m})R_A + (12\text{ m})(40\text{ m})\left(30\ \frac{\text{N}}{\text{m}}\right) \\ -(2\text{ m})(400\text{ N}) &= 0 \\ R_A &= 425\text{ N}\end{aligned}$$

Make a cut at point C, 10 m from the left support. Construct a free-body diagram of the section of beam to the left of the cut. (The right side could be used as an alternative.)

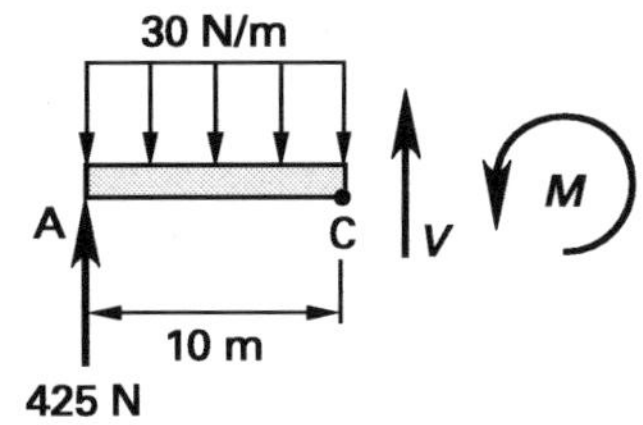

Perform a vertical force balance to determine the shear at point C.

$$\begin{aligned}\sum F_y &= 0 \\ 425\text{ N} - (10\text{ m})\left(30\ \frac{\text{N}}{\text{m}}\right) + V &= 0 \\ V &= -125\text{ N} \quad (130\text{ N})\end{aligned}$$

Whether the shear is positive or negative depends on the sign convention and free-body selected.

The answer is (A).

26. Calculate the moment of inertia.

$$\begin{aligned}I &= \frac{\pi}{64}(D^4 - d^4) = \frac{\pi}{64}\left((2\text{ in})^4 - (1.75\text{ in})^4\right) \\ &= 0.325\text{ in}^4\end{aligned}$$

Consider the shaft as a simply supported beam with two intermediate loads. Calculate deflections at A and B due to each load.

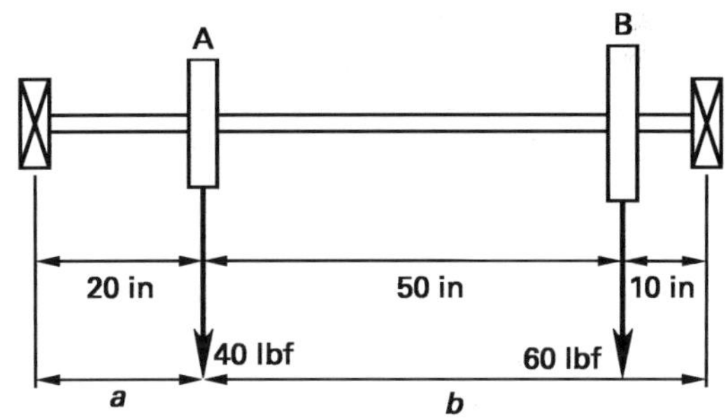

Calculate deflection at A due to the 40 lbf load.

$$\begin{aligned}y_{A,40} &= \frac{Pa^2b^2}{3EIL} \\ &= \frac{(40\text{ lbf})(20\text{ in})^2(60\text{ in})^2}{(3)\left(30.5\times 10^6\ \frac{\text{lbf}}{\text{in}^2}\right)(0.325\text{ in}^4)(80\text{ in})} \\ &= 0.024\text{ in}\end{aligned}$$

Calculate deflection at B due to the 40 lbf load.

$$y_{B,40} = \left(\frac{Pb}{6EIL}\right)\left(\frac{L}{b}(x-a)^3 + (L^2-b^2)x - x^3\right)$$

$$= \left(\frac{(40\text{ lbf})(60\text{ in})}{(6)\left(30.5\times10^6\ \frac{\text{lbf}}{\text{in}^2}\right)(0.325\text{ in}^4)(80\text{ in})}\right) \times \left(\begin{array}{l}\left(\frac{80\text{ in}}{60\text{ in}}\right)(70\text{ in} - 20\text{ in})^3 \\ \quad +((80\text{ in})^2 - (60\text{ in})^2) \\ \quad \times(70\text{ in}) - (70\text{ in})^3\end{array}\right)$$

$$= 0.0099\text{ in}\quad(0.010\text{ in})$$

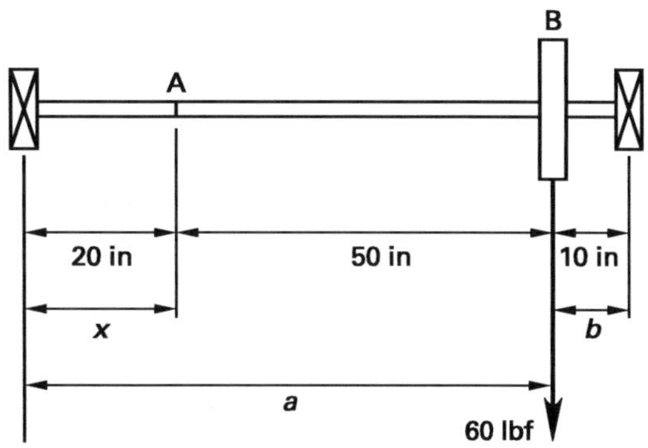

Calculate deflection at B due to the 60 lbf load.

$$y_{B,60} = \frac{Pa^2b^2}{3EIL}$$

$$= \frac{(60\text{ lbf})(70\text{ in})^2(10\text{ in})^2}{(3)\left(30.5\times10^6\ \frac{\text{lbf}}{\text{in}^2}\right)(0.325\text{ in}^4)(80\text{ in})}$$

$$= 0.012\text{ in}$$

Calculate deflection at A due to the 60 lbf load.

$$y_{A,60} = \left(\frac{Pb}{6EIL}\right)(L^2x - b^2x - x^3)$$

$$= \left(\frac{(60\text{ lbf})(10\text{ in})}{(6)\left(30.5\times10^6\ \frac{\text{lbf}}{\text{in}^2}\right)(0.325\text{ in}^4)(80\text{ in})}\right) \times \left(\begin{array}{l}(80\text{ in})^2(20\text{ in}) \\ \quad -(10\text{ in})^2(20\text{ in}) - (20\text{ in})^3\end{array}\right)$$

$$= 0.0149\text{ in}\quad(0.015\text{ in})$$

Calculate the total deflection at A and B.

$$\begin{aligned}y_{A,\text{total}} &= y_{A,40} + y_{A,60} \\ &= 0.024\text{ in} + 0.015\text{ in} \\ &= 0.039\text{ in} \\ y_{B,\text{total}} &= y_{B,40} + y_{B,60} \\ &= 0.010\text{ in} + 0.012\text{ in} \\ &= 0.022\text{ in}\end{aligned}$$

Using the Rayleigh-Ritz equation, calculate the shaft's lowest critical speed.

$$f = \frac{1}{2\pi}\sqrt{\frac{g(W_A y_{A,\text{total}} + W_B y_{B,\text{total}})}{W_A y_{A,\text{total}}^2 + W_B y_{B,\text{total}}^2}}$$

$$= \left(\frac{1\text{ rev}}{2\pi\text{ rad}}\right) \times\sqrt{\frac{\left(384\ \frac{\text{in}}{\text{sec}^2}\right)\left(\begin{array}{l}(40\text{ lbf})(0.039\text{ in}) \\ \quad +(60\text{ lbf})(0.022\text{ in})\end{array}\right)}{(40\text{ lbf})(0.039\text{ in})^2 + (60\text{ lbf})(0.022\text{ in})^2}} \times\left(60\ \frac{\text{sec}}{\text{min}}\right)$$

$$= 1059\text{ rpm}\quad(1100\text{ rpm})$$

The answer is (C).

27. The polar moment of inertia of the hollow shaft is

$$J = \frac{\pi D^4}{32}$$

Reformulate the expression for equivalent normal stress to solve for the inside diameter.

$$\sigma' = \frac{r}{J}\sqrt{4M^2 + 3T^2} = \frac{D}{2J}\sqrt{4M^2+3T^2}$$

$$= \left(\frac{16D}{\pi(D^4 - d^4)}\right)\sqrt{4M^2+3T^2}$$

$$d = \left(D^4 - \frac{16D}{\pi\sigma'}\sqrt{4M^2+3T^2}\right)^{1/4}$$

Restate the limiting stress using the maximum shear stress criterion.

$$\sigma' = \frac{S_y}{\text{FS}} = \frac{S_y}{2}$$

Solve for the maximum inside diameter.

$$d = \left(D^4 - \frac{16D(\text{FS})}{\pi S_y}\sqrt{4M^2 + 3T^2}\right)^{1/4}$$

$$= \left((3.0 \text{ in})^4 - \frac{(16)(3.0 \text{ in})(2)}{\pi\left(80{,}000 \frac{\text{lbf}}{\text{in}^2}\right)} \times \sqrt{\begin{array}{l}(4)(10{,}000 \text{ in-lbf})^2 \\ +(3)(15{,}000 \text{ in-lbf})^2\end{array}} \right)^{1/4}$$

$$= 2.88 \text{ in} \quad (2.9 \text{ in})$$

The answer is (D).

28. Calculate total shear area. The bolts are loaded in double shear.

$$A_s = (2 \text{ bolts})\left(2 \frac{\text{surfaces}}{\text{bolt}}\right)\left(\frac{\pi}{4}d^2\right) = \pi d^2$$

Using the distortion energy theory of failure, determine the ultimate shear strength.

$$S_{us} = 0.577S_u = (0.577)\left(74{,}500 \frac{\text{lbf}}{\text{in}^2}\right)$$
$$= 42{,}990 \text{ lbf/in}^2$$

Determine the force that the pattern can support with each bolt size.

$$F = S_{us}A_s$$

size (in)	S_{us} (lbf/in^2)	A_s (in^2)	force supported (lbf)
1/4	42,990	0.1963	8441
3/8	**42,990**	**0.4418**	**18,992**
1/2	42,990	0.7854	33,764
5/8	42,990	1.2272	52,757

The answer is (B).

29. Calculate the bolt force at the proof stress.

$$F = \sigma_{\text{proof}}A_t = \left(105{,}000 \frac{\text{lbf}}{\text{in}^2}\right)(0.606 \text{ in}^2)$$
$$= 63{,}630 \text{ lbf}$$

Determine the torque required to tighten the nut.

$$\alpha = 30°$$
$$\theta = 2.48°$$
$$r_t = \frac{d_t}{2} = \frac{0.9188 \text{ in}}{2} = 0.4594 \text{ in}$$
$$T = \left(\frac{f_c r_c}{d_{\text{bolt}}} + \left(\frac{r_t}{d_{\text{bolt}}}\right)\left(\frac{\tan\theta + f_t \sec\alpha}{1 - f_t \tan\theta \sec\alpha}\right)\right) d_{\text{bolt}}F$$

$$= \frac{\left(\begin{array}{l}\dfrac{(0.15)(0.625 \text{ in})}{1.00 \text{ in}} \\ +\left(\dfrac{0.4594 \text{ in}}{1.00 \text{ in}}\right)\left(\dfrac{\tan 2.48° + 0.15 \sec 30°}{1 - 0.15 \tan 2.48° \sec 30°}\right)\end{array}\right) \times (1.00 \text{ in})(63{,}630 \text{ lbf})}{12 \frac{\text{in}}{\text{ft}}}$$

$$= 1029 \text{ ft-lbf} \quad (1000 \text{ ft-lbf})$$

The answer is (A).

30. The problem can be solved analytically using the Freudenstein equation, a fundamental kinematic identity.

$$R_1 \cos\phi - R_2 \cos\theta + R_3 = \cos(\phi - \theta)$$

Simplify using a trigonometric identity, and solve for the coefficients.

$$\cos(\phi - \theta) = \cos\phi\cos\theta + \sin\phi\sin\theta$$
$$R_1 = \frac{a}{d} = \frac{5 \text{ in}}{4 \text{ in}} = 5/4$$
$$R_2 = \frac{a}{b} = \frac{5 \text{ in}}{2 \text{ in}} = 5/2$$
$$R_3 = \frac{b^2 - c^2 + d^2 + a^2}{2bd}$$
$$= \frac{(2 \text{ in})^2 - (3 \text{ in})^2 + (4 \text{ in})^2 + (5 \text{ in})^2}{(2)(2 \text{ in})(4 \text{ in})}$$
$$= 36/16$$

Substituting and simplifying,

$$R_1 \cos\phi - R_2 \cos\theta + R_3 = \cos\phi\cos\theta + \sin\phi\sin\theta$$
$$\frac{5}{4}\cos 120° - \frac{5}{2}\cos\theta + \frac{36}{16} = \cos 120° \cos\theta + \sin 120° \sin\theta$$
$$1 = \frac{16}{13}\cos\theta + \frac{4\sqrt{3}}{13}\sin\theta$$

Solve for the output angle by selecting trial θ values and tabulating results.

θ (deg)	$1-\frac{4\sqrt{3}}{13}\sin\theta$	$\frac{16}{13}\cos\theta$
55	0.563	0.706
65	0.517	0.520
65.2 (65)	**0.516**	**0.516**
75	0.485	0.318
85	0.469	0.107

The answer is (B).

31. Find the maximum (larger) principal stress.

$$\begin{aligned}\sigma_1 &= \frac{\sigma_x+\sigma_y}{2}+\sqrt{\left(\frac{\sigma_x-\sigma_y}{2}\right)^2+\tau_{xy}^2}\\ &= \frac{10{,}000\ \frac{\text{lbf}}{\text{in}^2}+15{,}000\ \frac{\text{lbf}}{\text{in}^2}}{2}\\ &\quad+\sqrt{\left(\frac{10{,}000\ \frac{\text{lbf}}{\text{in}^2}-15{,}000\ \frac{\text{lbf}}{\text{in}^2}}{2}\right)^2+\left(5000\ \frac{\text{lbf}}{\text{in}^2}\right)^2}\\ &= 18{,}090\ \text{lbf/in}^2\end{aligned}$$

Find the maximum shear stress.

$$\begin{aligned}\tau_{\text{max}} &= \sqrt{\left(\frac{\sigma_x-\sigma_y}{2}\right)^2+\tau_{xy}^2}\\ &= \sqrt{\left(\frac{10{,}000\ \frac{\text{lbf}}{\text{in}^2}-15{,}000\ \frac{\text{lbf}}{\text{in}^2}}{2}\right)^2+\left(5000\ \frac{\text{lbf}}{\text{in}^2}\right)^2}\\ &= 5590\ \text{lbf/in}^2\end{aligned}$$

Find the factor of safety for normal stress.

$$\begin{aligned}\text{FS} &= \frac{S_{ut}}{\sigma_1} = \frac{65{,}000\ \frac{\text{lbf}}{\text{in}^2}}{18{,}090\ \frac{\text{lbf}}{\text{in}^2}}\\ &= 3.59\quad (3.6)\end{aligned}$$

Find the factor of safety for shear stress.

$$\begin{aligned}\text{FS} &= \frac{\tau_{\text{yield}}}{\tau_{\text{max}}} = \frac{10{,}000\ \frac{\text{lbf}}{\text{in}^2}}{5590\ \frac{\text{lbf}}{\text{in}^2}}\\ &= 1.79\quad (1.8)\quad \text{[controls]}\end{aligned}$$

The answer is (A).

32. Calculate the required thermal expansion coefficient.

$$\begin{aligned}\frac{x_2-x_1}{(T_2-T_1)x_1} &= \frac{10.045\ \text{in}-10.000\ \text{in}}{(80°\text{C}-25°\text{C})(10.000\ \text{in})}\\ &= 81.8\times 10^{-6}\ \text{in/in-°C}\end{aligned}$$

Material I has a higher coefficient of thermal expansion and is disqualified. Calculate the required specific gravity.

$$\text{SG}_{\text{req}} \le 0.5(\text{SG}_{\text{al}}) \le (0.5)(2.71) = 1.355$$

Material II has a higher specific gravity than required and is disqualified. Calculate the operating temperature in degrees Fahrenheit.

$$T_{°\text{F}} = \tfrac{9}{5}T_{°\text{C}}+32° = \left(\frac{9}{5}\right)(80°\text{C})+32° = 176°\text{F}$$

Material III has a lower maximum temperature and is disqualified. Material IV is the only structural polymer of this group that meets the design engineers' requirements.

The answer is (D).

33. Solve using the equivalent uniform annual cost method.

Calculate the annual cost of the upgrade alternative.

$$\begin{aligned}\text{EUAC}_u &= \text{annual maintenance}\\ &\quad+(\text{initial cost}+\text{present salvage value})\\ &\quad\times(\text{capital recovery factor})\\ &\quad-(\text{future salvage value})\\ &\quad\times(\text{uniform series sinking fund factor})\\ &= \$500+(\$9000+\$13{,}000)(A/P,8\%,20)\\ &\quad-(\$10{,}000)(A/F,8\%,20)\\ &= \$500+(\$9000+\$13{,}000)(0.1019)\\ &\quad-(\$10{,}000)(0.0219)\\ &= \$2523\end{aligned}$$

Calculate the annual cost of the replacement alternative.

$$\begin{aligned}\text{EUAC}_r &= \text{annual maintenance}\\ &\quad +(\text{initial cost})(\text{capital recovery factor})\\ &\quad -(\text{future salvage value})\\ &\quad \times(\text{uniform series sinking fund factor})\\ &= \$100 + (\$40{,}000)(A/P, 8\%, 25)\\ &\quad -(\$15{,}000)(A/F, 8\%, 25)\\ &= \$100 + (\$40{,}000)(0.0937)\\ &\quad -(\$15{,}000)(0.0137)\\ &= \$3642\end{aligned}$$

The difference in annual costs is

$$\begin{aligned}\text{EUAC}_r - \text{EUAC}_u &= \$3642 - \$2523\\ &= \$1119 \quad (\$1100)\end{aligned}$$

The upgrade alternative is more economical.

The answer is (C).

34. Radial deformation of the shaft during rotation is neglected.

For aluminum,

$$\begin{aligned}E &= 10 \times 10^6 \text{ psi}\\ \mu &= 0.33\\ \rho &= \frac{173 \ \frac{\text{lbm}}{\text{ft}^3}}{\left(12 \ \frac{\text{in}}{\text{ft}}\right)^3} = 0.100 \text{ lbm/in}^3\\ \gamma &= 0.100 \text{ lbf/in}^3\end{aligned}$$

For steel,

$$\begin{aligned}E &= 2.9 \times 10^7 \text{ psi}\\ \mu &= 0.30\end{aligned}$$

The tangential stress is

$$\sigma_t = E\epsilon_i = E\left(\frac{\Delta r_i}{r_i}\right)$$

Solving for Δr_i,

$$\Delta r_i = \frac{r_i\sigma_t}{E}$$

$$r_o = \frac{12 \text{ in}}{2} = 6 \text{ in}$$

$$r_i = \frac{2 \text{ in}}{2} = 1 \text{ in}$$

Calculate σ_t ($g = 386.4$ in/sec^2). The tangential stress is maximum at the inner face.

$$\omega = 2\pi f = 2\pi\left(\frac{n}{60 \ \frac{\text{sec}}{\text{min}}}\right)$$

$$\Delta r_i = \left(\frac{1}{4}\right)\left(\frac{\gamma\omega^2 r_i}{386.4E}\right)\left((3+\mu)r_o^2 + (1-\mu)r_i^2\right)$$

$$\frac{0.0045 \text{ in}}{2} = \left(\frac{1}{4}\right)\left(\frac{\left(0.100 \ \frac{\text{lbf}}{\text{in}^2}\right)\left(2\pi\left(\frac{n}{60 \ \frac{\text{sec}}{\text{min}}}\right)\right)^2 (1 \text{ in})}{\left(386.4 \ \frac{\text{in}}{\text{sec}^2}\right)\left(10 \times 10^6 \ \frac{\text{lbf}}{\text{in}^2}\right)}\right) \times\left((3+0.33)(6 \text{ in})^2 + (1-0.33)(1 \text{ in})^2\right)$$

$$n = 16{,}219 \text{ rpm} \quad (1.6 \times 10^4 \text{ rpm})$$

The answer is (A).

35. Tabulate fatigue life from the endurance chart.

completely reversed stress (lbf/in^2)	fatigue life (cycles)
±80,000	5×10^4
±50,000	3×10^5
±30,000	1×10^6

Apply the linear cumulative damage (Palmgren-Miner) rule.

$$\sum_{j=1}^{k} \frac{n_j}{N_j} = 1$$

Calculate the lifetime fraction consumed during one 9 min period.

$$\begin{aligned}L_{\text{fraction,1 period}} &= \frac{1}{5 \times 10^4} + \frac{1}{3 \times 10^5} + \frac{1}{1 \times 10^6}\\ &= 2.43 \times 10^{-5} \text{ lifetime}\end{aligned}$$

Calculate the total number of periods per lifetime.

$$\begin{aligned} N_{\text{total}} &= \frac{1}{L_{\text{fraction,1 period}}} \\ &= \frac{1 \text{ period}}{2.43 \times 10^{-5} \text{ lifetime}} \\ &= 41{,}152 \text{ periods/lifetime} \end{aligned}$$

Calculate the lifetime in hours.

$$\begin{aligned} t_{\text{total}} &= \left(41{,}152 \ \frac{\text{periods}}{\text{lifetime}}\right)\left(\frac{9 \ \frac{\text{min}}{\text{period}}}{60 \ \frac{\text{min}}{\text{hr}}}\right) \\ &= 6173 \text{ hr/lifetime} \quad (6200 \text{ hr}) \end{aligned}$$

The answer is (B).

36. Calculate the effective length. For fixed-free end conditions

$$\begin{aligned} C &= 2.1 \\ L_e &= CL = (2.1)(2 \text{ in}) \\ &= 4.2 \text{ in} \end{aligned}$$

Calculate the section area, minimum moment of inertia, minimum radius of gyration, and slenderness ratio.

$$\begin{aligned} A &= bh = (1.0 \text{ in})(0.7 \text{ in}) = 0.7 \text{ in}^2 \\ I_{\text{min}} &= \frac{1}{12}bh^3 = \left(\frac{1}{12}\right)(1.0 \text{ in})(0.7 \text{ in})^3 = 0.0286 \text{ in}^4 \\ r_{\text{min}} &= \sqrt{\frac{I_{\text{min}}}{A}} = \sqrt{\frac{0.0286 \text{ in}^4}{0.7 \text{ in}^2}} = 0.2 \text{ in} \\ \text{SR} &= \frac{L_e}{r_{\text{min}}} = \frac{4.2 \text{ in}}{0.2 \text{ in}} = 21 \end{aligned}$$

Calculate the critical slenderness ratio. If the slenderness ratio is less than the critical slenderness ratio, the column is not long, and the Johnson formula is used to calculate the critical load.

$$\begin{aligned} \text{SR}_{\text{critical}} &= \sqrt{\frac{2\pi^2 E}{S_y}} = \sqrt{\frac{2\pi^2\left(420{,}000 \ \frac{\text{lbf}}{\text{in}^2}\right)}{10{,}000 \ \frac{\text{lbf}}{\text{in}^2}}} \\ &= 28.8 \end{aligned}$$

Since $21 < 28.8$, use the Johnson formula to calculate the critical load.

$$\begin{aligned} P_{\text{critical}} &= AS_y\left(1 - \frac{S_y(\text{SR})^2}{4\pi^2 E}\right) \\ &= (0.7 \text{ in}^2)\left(10{,}000 \ \frac{\text{lbf}}{\text{in}^2}\right) \\ &\quad \times \left(1 - \frac{\left(10{,}000 \ \frac{\text{lbf}}{\text{in}^2}\right)(21)^2}{4\pi^2\left(420{,}000 \ \frac{\text{lbf}}{\text{in}^2}\right)}\right) \\ &= 5138 \text{ lbf} \quad (5100 \text{ lbf}) \end{aligned}$$

The answer is (A).

37. For processes conforming to the standard normal distribution, probability of a given outcome is calculated using the integral of the standard normal distribution equation. The standard normal distribution equation is

$$f(x) = \frac{1}{\sqrt{2\pi}}e^{-\frac{1}{2}x^2}$$

$$\int_{-\infty}^{+\infty} f(x)\,dx = 1$$

Integration across the entire domain equals π as shown, but the solution for intermediate domains is difficult. Statistics texts provide tabulated solutions. The z-transform simplifies the probability calculation by normalizing standard deviations and setting the mean to zero. The z-transform curve is shown.

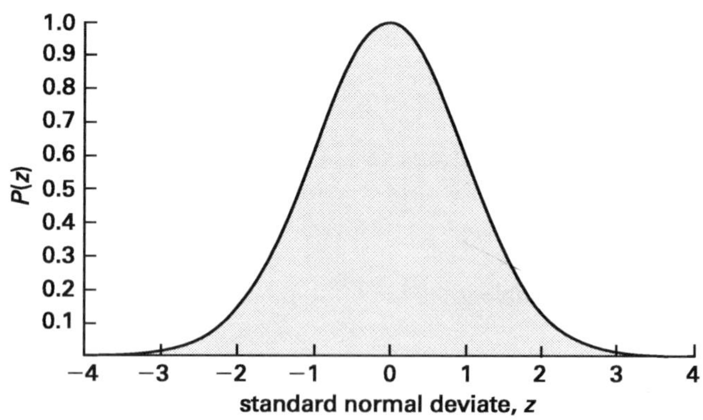

The table shown provides the z-transform solution for integer values of z.

	A	B	C	D	E
z	area $-\infty$ to z	area z to $+\infty$	A + B reject	1 − C accept	percent
0.00	0.5000	0.5000	1.0000	0.0000	0.0%
1.00	0.1587	0.1587	0.3174	0.6826	68.3%
2.00	0.0228	0.0228	0.0456	0.9544	95.4%
3.00	0.0014	0.0014	0.0028	0.9972	99.7%
4.00	3.2×10^{-5}	3.2×10^{-5}	6.4×10^{-5}	0.9999	99.9%

The probability of an accepted process outcome is represented by column D. For example, for $-1 < z < 1$ (corresponding to $\pm 1\sigma$) the probability of success is 0.6826. For $-2 < z < 2$ ($\pm 2\sigma$) the probability of success is 0.9544. For $-3 < z < 3$ ($\pm 3\sigma$) the probability of success is 0.9972 (99.7%).

The answer is (C).

38. Evaluate each joint, calculating the effective allowable stress, geometry after corrosion, and allowable pressure. Verify that the results meet code limitations. Evaluate the type 1 spot radiographed longitudinal joint. Calculate the effective allowable stress.

$$\begin{aligned}\sigma_{850°F} &= 8.7 \times 10^3 \text{ lbf/in}^2 \\ SE_{\text{long}} &= \sigma_{850°F}E \\ &= \left(8.7 \times 10^3 \frac{\text{lbf}}{\text{in}^2}\right)(0.85) \\ &= 7395 \text{ lbf/in}^2\end{aligned}$$

Calculate the relevant geometry (shell thickness and inside radius) after corrosion.

$$\begin{aligned}t_c &= t - \text{CA} = 0.625 \text{ in} - \frac{1}{16} \text{ in} \\ &= 0.5625 \text{ in} \\ R_c &= \frac{\text{OD}}{2} - t + \text{CA} = \frac{24 \text{ in}}{2} - 0.625 \text{ in} + \frac{1}{16} \text{ in} \\ &= 11.44 \text{ in}\end{aligned}$$

Calculate the maximum allowed pressure.

$$\begin{aligned}p_{\max} &= \frac{SE_{\text{long}}t_c}{R_c + 0.6t_c} \\ &= \frac{\left(7395 \frac{\text{lbf}}{\text{in}^2}\right)(0.5625 \text{ in})}{11.44 \text{ in} + (0.6)(0.5625 \text{ in})} \\ &= 353 \text{ lbf/in}^2\end{aligned}$$

Verify that results meet code limitations.

$$\begin{aligned}p_{\max} &= 353 \frac{\text{lbf}}{\text{in}^2} \le 0.385SE \\ &= 353 \frac{\text{lbf}}{\text{in}^2} \le (0.385)\left(7395 \frac{\text{lbf}}{\text{in}^2}\right) \\ &= 353 \text{ lbf/in}^2 \le 2847 \text{ lbf/in}^2 \quad \text{[OK]} \\ t_c &= 0.5625 \text{ in} \le 0.5R_c \\ &= 0.5625 \text{ in} \le (0.5)(11.44 \text{ in}) \\ &= 0.5625 \text{ in} \le 5.72 \text{ in} \quad \text{[OK]}\end{aligned}$$

Evaluate the type 2 spot radiographed circumferential joint. Calculate the effective allowable stress.

$$\begin{aligned}SE_{\text{circ}} &= \sigma_{850°F}E = \left(8.7 \times 10^3 \frac{\text{lbf}}{\text{in}^2}\right)(0.80) \\ &= 6960 \text{ lbf/in}^2\end{aligned}$$

Calculate the relevant geometry after corrosion. (Use the shell thickness and inside radius calculated for the longitudinal joint.)

$$\begin{aligned}t_c &= 0.5625 \text{ in} \\ R_c &= 11.44 \text{ in}\end{aligned}$$

Calculate the maximum allowable pressure.

$$\begin{aligned}p_{\max} &= \frac{2SE_{\text{circ}}t_c}{R_c - 0.4t_c} \\ &= \frac{(2)\left(6960 \frac{\text{lbf}}{\text{in}^2}\right)(0.5625 \text{ in})}{11.44 \text{ in} - (0.4)(0.5625 \text{ in})} \\ &= 698 \text{ lbf/in}^2\end{aligned}$$

Verify that results meet code limitations.

$$\begin{aligned}p_{\max} &= 698 \frac{\text{lbf}}{\text{in}^2} \le 1.25SE \\ &= 698 \frac{\text{lbf}}{\text{in}^2} \le (1.25)\left(6960 \frac{\text{lbf}}{\text{in}^2}\right) \\ &= 698 \text{ lbf/in}^2 \le 8700 \text{ lbf/in}^2 \quad \text{[OK]} \\ t_c &= 0.5625 \text{ in} \le 0.5R_c \\ &= 0.5625 \text{ in} \le (0.5)(11.44 \text{ in}) \\ &= 0.5625 \text{ in} \le 5.72 \text{ in} \quad \text{[OK]}\end{aligned}$$

Evaluate the elliptical head. Calculate the effective allowable stress.

$$\begin{aligned} SE_{\text{ehead}} &= \sigma_{850^\circ\text{F}}E \\ &= \left(8.7\times 10^3\ \frac{\text{lbf}}{\text{in}^2}\right)(1.0) \\ &= 8700\ \text{lbf/in}^2 \end{aligned}$$

Calculate the relevant geometry (shell thickness, shell diameter, and K-factor) after corrosion. (Use the shell thickness calculated for the longitudinal joint.)

$$\begin{aligned} t_c &= 0.5625\ \text{in} \\ D_c &= 2R_c = (2)(11.44\ \text{in}) = 22.88\ \text{in} \\ h &= 6\ \text{in} \\ K &= \left(\frac{1}{6}\right)\left(2+\left(\frac{D_c}{2h}\right)^2\right) = \left(\frac{1}{6}\right)\left(2+\left(\frac{22.88\ \text{in}}{(2)(6\ \text{in})}\right)^2\right) \\ &= 0.939 \end{aligned}$$

Calculate maximum allowable pressure.

$$\begin{aligned} p_{\max} &= \frac{2SE_{\text{ehead}}t_c}{KD_c+0.2t_c} \\ &= \frac{(2)\left(8700\ \frac{\text{lbf}}{\text{in}^2}\right)(0.5625\ \text{in})}{(0.939)(22.88\ \text{in})+(0.2)(0.5625\ \text{in})} \\ &= 453\ \text{lbf/in}^2 \end{aligned}$$

Code limitations were satisfied by the calculation of K.

Evaluate the hemispherical head. Calculate the effective allowable stress.

$$\begin{aligned} SE_{\text{hhead}} &= \sigma_{850^\circ\text{F}}E \\ &= \left(8.7\times 10^3\ \frac{\text{lbf}}{\text{in}^2}\right)(0.80) \\ &= 6960\ \text{lbf/in}^2 \end{aligned}$$

Calculate the relevant geometry (head thickness and inside radius) after corrosion.

$$\begin{aligned} t_c &= t - \text{CA} = 0.5\ \text{in} - \frac{1}{16}\ \text{in} \\ &= 0.4375\ \text{in} \\ R_c &= R - t - \text{CA} = \frac{24\ \text{in}}{2} - 0.5\ \text{in} - \frac{1}{16}\ \text{in} \\ &= 11.438\ \text{in} \end{aligned}$$

Calculate the maximum allowable pressure.

$$\begin{aligned} p_{\max} &= \frac{2SE_{\text{hhead}}t_c}{R_c+0.2t_c} \\ &= \frac{(2)\left(6960\ \frac{\text{lbf}}{\text{in}^2}\right)(0.4375\ \text{in})}{11.438\ \text{in}+(0.2)(0.4375\ \text{in})} \\ &= 528\ \text{lbf/in}^2 \end{aligned}$$

Verify that results meet code limitations.

$$\begin{aligned} p_{\max} &= 528\ \frac{\text{lbf}}{\text{in}^2} \le 0.665SE \\ &= 528\ \frac{\text{lbf}}{\text{in}^2} \le (0.665)\left(6960\ \frac{\text{lbf}}{\text{in}^2}\right) \\ &= 528\ \text{lbf/in}^2 \le 4628\ \text{lbf/in}^2 \quad \text{[OK]} \\ t_c &= 0.4375\ \text{in} \le 0.356R_c \\ &= 0.4375\ \text{in} \le (0.356)(11.438\ \text{in}) \\ &= 0.4375\ \text{in} \le 4.07\ \text{in} \quad \text{[OK]} \end{aligned}$$

The lowest allowable stress, 353 lbf/in^2 (350 lbf/in^2), occurs on the longitudinal weld.

The answer is (A).

39. Calculate the impact factor.

$$\begin{aligned} f_i &= 1+\sqrt{1+\frac{2h}{\delta_{st}}} \\ &= 1+\sqrt{1+\frac{(2)(6\ \text{ft})\left(12\ \frac{\text{in}}{\text{ft}}\right)}{5\times 10^{-3}\ \text{in}}} \\ &= 170.7 \end{aligned}$$

Use the impact factor, effective force, failure deflection, and factor of safety to find the allowable weight.

$$\begin{aligned} \frac{F_e}{N} &= Wf_i \\ W &= \frac{F_e}{Nf_i} = \frac{1250\ \text{lbf}}{(6)(170.7)} \\ &= 1.22\ \text{lbf} \quad (1.2\ \text{lbf}) \end{aligned}$$

The answer is (D).

40. The cycloidal profile provides zero acceleration at the start of rise and end of return. The harmonic and parabolic profiles impose instantaneous accelerations upon the start of rise and end of return. There is no profile named the velocity derivative.

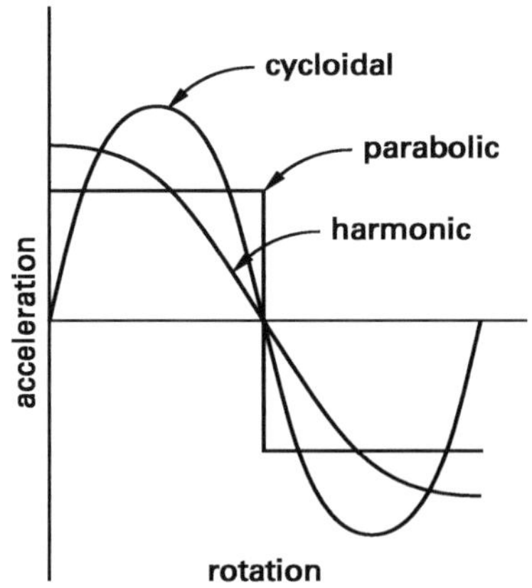

The answer is (B).

41. The thickness of the beam is doubled. The thickness ratio, t_2/t_1, is 2. The cross section is rectangular, so the mass ratio, m_2/m_1, is also 2. The moment of inertia ratio is

$$\begin{aligned}\frac{I_2}{I_1} &= \frac{\frac{1}{12}bt_2^3}{\frac{1}{12}bt_1^3} = \left(\frac{t_2}{t_1}\right)^3 = (2)^3 \\ &= 8\end{aligned}$$

The stiffness constant for a cantilever beam is

$$k = \frac{P}{\delta} = \frac{3EI}{L^3}$$

The stiffness ratio is

$$\begin{aligned}\frac{k_2}{k_1} &= \frac{\frac{3EI_2}{L^3}}{\frac{3EI_1}{L^3}} = \frac{I_2}{I_1} \\ &= 8\end{aligned}$$

The fundamental natural frequency ratio is

$$\begin{aligned}\frac{\omega_{n,2}}{\omega_{n,1}} &= \frac{\sqrt{\frac{k_2}{m_2}}}{\sqrt{\frac{k_1}{m_1}}} = \sqrt{\frac{\frac{k_2}{k_1}}{\frac{m_2}{m_1}}} = \sqrt{\frac{8}{2}} \\ &= 2\end{aligned}$$

The answer is (C).

42. Although a third party often assists in design, evaluation, testing, and regulatory interpretation, the CE marking is not a third-party approval mark. It is a self-declaration under the supplier's own responsibility that the product conforms to European Union requirements. By affixing the CE marking to a product, the responsible company official declares successful completion of appropriate conformity assessment procedures.

The answer is (B).

43. The moment about a fulcrum, such as the end of a beam, is the cross product of the force applied and its associated position vector. When multiple forces exist, the net moment is the sum of all cross products.

The moment created by the distributed load is

$$\begin{aligned}M_{\text{distributed}} &= Fd = wL\left(d + \frac{L}{2}\right) \\ &= \left(20\ \frac{\text{lbf}}{\text{ft}}\right)(8\ \text{ft})\left(7\ \text{ft} + \frac{8\ \text{ft}}{2}\right) \\ &= 1760\ \text{ft-lbf}\end{aligned}$$

The moment created by the concentrated load is

$$\begin{aligned}M_{\text{concentrated}} &= Fd = (100\ \text{lbf})(15\ \text{ft}) \\ &= 1500\ \text{ft-lbf}\end{aligned}$$

The net moment is

$$\begin{aligned}M_{\text{net}} &= M_{\text{distributed}} + M_{\text{concentrated}} \\ &= 1760\ \text{ft-lbf} + 1500\ \text{ft-lbf} \\ &= 3260\ \text{ft-lbf} \quad (3300\ \text{ft-lbf})\end{aligned}$$

The answer is (C).

44. The moment of inertia for the rod is

$$\begin{aligned}I &= \frac{\pi}{64}d^4 = \frac{\pi}{64}(2\ \text{in})^4 \\ &= 0.7854\ \text{in}^4\end{aligned}$$

The modulus of elasticity for steel is 30×10^6 psi. The stiffness constant for a cantilever beam is

$$\begin{aligned}k &= \frac{P}{\delta} = \frac{3EI}{L^3} \\ &= \frac{(3)\left(30 \times 10^6\ \frac{\text{lbf}}{\text{in}^2}\right)(0.7854\ \text{in}^4)}{(36\ \text{in})^3} \\ &= 1515\ \text{lbf/in}\end{aligned}$$

The natural frequency is

$$\omega_n = \sqrt{\frac{kg_c}{m}} = \sqrt{\frac{\left(1515\ \frac{\text{lbf}}{\text{in}}\right)\left(32.2\ \frac{\text{lbm-ft}}{\text{lbf-sec}^2}\right)\left(12\ \frac{\text{in}}{\text{ft}}\right)}{50\ \text{lbm}}} = 108.2\ \text{rad/sec}$$

The damping frequency is

$$\zeta = \frac{Cg_c}{2m\omega_n} = \frac{\left(1.0\ \frac{\text{lbf-sec}}{\text{in}}\right)\left(32.2\ \frac{\text{lbm-ft}}{\text{lbf-sec}^2}\right)\left(12\ \frac{\text{in}}{\text{ft}}\right)}{(2)\left(50\ \text{lbm}\right)\left(108.2\ \frac{\text{rad}}{\text{sec}}\right)} = 0.0357\quad (0.036)$$

The answer is (B).

45. Draw a free-body diagram for the given dynamic conditions. Note that the pulley system doubles the pull on the crate toward the top of the ramp.

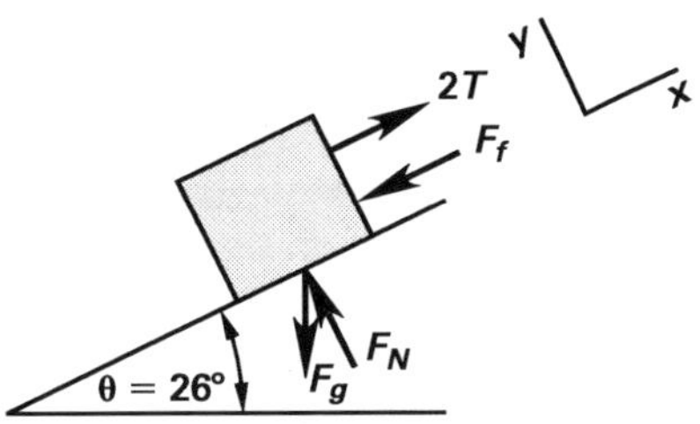

Determine the normal force acting on the crate by using a force balance in the direction perpendicular to the plane of motion (y-direction).

$$\sum F_y = 0$$
$$F_N - F_g\cos\theta = 0$$
$$F_N = F_g\cos\theta = mg\cos\theta = (500\ \text{kg})\left(9.81\ \frac{\text{m}}{\text{s}^2}\right)\cos 26^\circ = 4409\ \text{N}$$

Because the velocity of the tractor is constant, the crate's acceleration up the ramp is zero. A new force balance in the direction of motion (x-direction) yields

$$\sum F_x = ma_x = 0$$
$$2T - F_f - F_g\sin\theta = 0$$

Solve for the tension in the cable, T.

$$T = \tfrac{1}{2}(F_f + F_g\sin\theta) = \tfrac{1}{2}(\mu_k F_N + mg\sin\theta) = \left(\frac{1}{2}\right)\left((0.24)(4409\ \text{N}) + (500\ \text{kg})\left(9.81\ \frac{\text{m}}{\text{s}^2}\right)\sin 26^\circ\right) = 1604\ \text{N}\quad (1600\ \text{N})$$

The answer is (C).

46. Cathodic protection is a technique to prevent metal surface corrosion by making that surface the cathode of an electrochemical cell. For example, surfaces can be cathodically protected through the application of either galvanic anodes or impressed current.

Anodic protection is the application of direct current to shift corrosion potential to the passive zone in active-passive metals exposed to strong alkaline or acidic environments.

Passivation is the process of cleaning metal surfaces and forming a corrosion-resistant film, particularly on stainless steel, but also on aluminum, copper, and other metals. For stainless steel, the surface removal of iron or iron compounds is accomplished by means of a chemical dissolution, typically an acid treatment. Treatment with a mild nitric acid solution or other oxidant forms a protective passive film.

In the process of galvanization, the metal surface is coated with protective zinc by dipping or electrodeposition.

The answer is (A).

47. The average force during each of the phases of operation is

$$\overline{F}_{0-5} = \frac{0\ \text{lbf} + 32\ \text{lbf}}{2} = 16\ \text{lbf}$$
$$\overline{F}_{5-9} = \frac{32\ \text{lbf} + 20\ \text{lbf}}{2} = 26\ \text{lbf}$$
$$\overline{F}_{9-11} = \frac{20\ \text{lbf} + 0\ \text{lbf}}{2} = 10\ \text{lbf}$$
$$\overline{F}_{11-16} = \overline{F}_{0-5} = 16\ \text{lbf}$$
$$\overline{F}_{16-20} = \overline{F}_{5-9} = 26\ \text{lbf}$$

The average force during the first 20 sec of operation is

$$\overline{F}_{0-20} = \frac{\overline{F}_{0-5}t_{0-5} + \overline{F}_{5-9}t_{5-9} + \overline{F}_{9-11}t_{9-11} + \overline{F}_{11-16}t_{11-16} + \overline{F}_{16-20}t_{16-20}}{t_{0-20}}$$

$$= \frac{(16 \text{ lbf})(5 \text{ sec}) + (26 \text{ lbf})(4 \text{ sec}) + (10 \text{ lbf})(2 \text{ sec}) + (16 \text{ lbf})(5 \text{ sec}) + (26 \text{ lbf})(4 \text{ sec})}{20 \text{ sec}}$$

$$= 19.4 \text{ lbf} \quad (19 \text{ lbf})$$

The answer is (D).

48. The wheel rotates as shown.

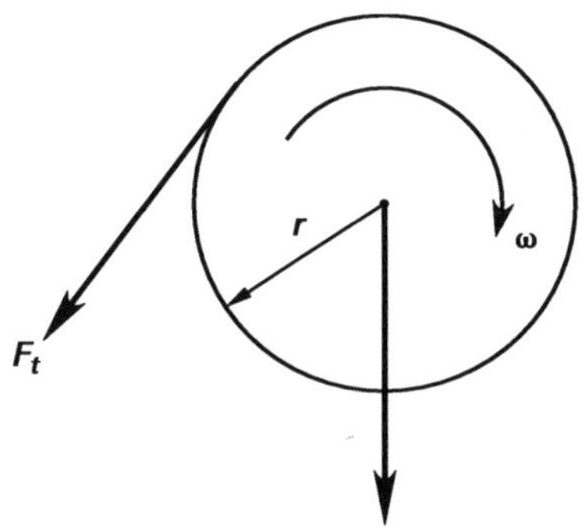

The torque, moment of inertia, angular velocity, and time are related by the equation

$$T = \left(\frac{I}{g_c}\right)\left(\frac{\Delta\omega}{\Delta t}\right)$$

The torque is also equal to the tangential force times the radius.

$$T = F_t r$$

The wheel is a solid cylinder, so its moment of inertia is

$$I = \frac{mr^2}{2}$$

Substituting into the first equation and solving for the tangential force gives

$$F_t r = \left(\frac{mr^2}{2g_c}\right)\left(\frac{\Delta\omega}{\Delta t}\right)$$

$$F_t = \frac{mr\Delta\omega}{2g_c\Delta t}$$

Reducing the speed of 54 rad/sec by a third means a change of 18 rad/sec. The tangential force needed to alter the angular velocity by 18 rad/sec in 30 sec is

$$F_t = \frac{mr\Delta\omega}{2g_c\Delta t}$$

$$= \frac{(90 \text{ lbm})(5 \text{ ft})\left(18 \ \frac{\text{rad}}{\text{sec}}\right)}{(2)\left(32.2 \ \frac{\text{lbm-ft}}{\text{lbf-sec}^2}\right)(30 \text{ sec})}$$

$$= 4.19 \text{ lbf} \quad (4 \text{ lbf})$$

The answer is (A).

49. The radius of the cylinder in feet is

$$r = \frac{15 \text{ in}}{12 \ \frac{\text{in}}{\text{ft}}} = 1.25 \text{ ft}$$

The mass moment of inertia of the cylinder is

$$I_{\text{cylinder}} = \tfrac{1}{2}m_{\text{cylinder}}r^2 = \left(\frac{1}{2}\right)(60 \text{ lbm})(1.25 \text{ ft})^2$$

$$= 46.88 \text{ lbm-ft}^2$$

The moment, M, equals the mass moment of inertia times the angular acceleration, α.

$$M = \frac{I_{\text{cylinder}}\alpha}{g_c}$$

$$rT = \left(\frac{I_{\text{cylinder}}}{g_c}\right)\left(\frac{a}{r}\right)$$

a is linear acceleration. The tension is

$$T = \frac{I_{\text{cylinder}}a}{g_c r^2} = \frac{(46.88 \text{ lbm-ft}^2)a}{\left(32.2 \ \frac{\text{lbm-ft}}{\text{lbf-sec}^2}\right)(1.25 \text{ ft})^2}$$

$$= \left(0.932 \ \frac{\text{lbf-sec}^2}{\text{ft}}\right)a$$

Use a force balance around the 18 lbm weight to find another relationship between tension and acceleration.

$$\begin{aligned}
\sum F &= \frac{m_{\text{weight}}a}{g_c} \\
F_{\text{weight}} - F_{\text{tension}} &= \frac{m_{\text{weight}}a}{g_c} \\
\frac{m_{\text{weight}}g}{g_c} - T &= \frac{m_{\text{weight}}a}{g_c} \\
T &= \frac{m_{\text{weight}}g}{g_c} - \left(\frac{m_{\text{weight}}}{g_c}\right)a \\
&= \frac{(18\text{ lbm})\left(32.2\ \frac{\text{ft}}{\text{sec}^2}\right)}{32.2\ \frac{\text{lbm-ft}}{\text{lbf-sec}^2}} - \left(\frac{18\text{ lbm}}{32.2\ \frac{\text{lbm-ft}}{\text{lbf-sec}^2}}\right)a \\
&= 18\text{ lbf} - \left(0.559\ \frac{\text{lbm-sec}^2}{\text{ft}}\right)a
\end{aligned}$$

Solve the two equations simultaneously to find the acceleration.

$$\begin{aligned}
T &= \left(0.932\ \frac{\text{lbf-sec}^2}{\text{ft}}\right)a \\
T &= 18\text{ lbf} - \left(0.559\ \frac{\text{lbf-sec}^2}{\text{ft}}\right)a \\
18\text{ lbf} &= \left(1.491\ \frac{\text{lbf-sec}^2}{\text{ft}}\right)a \\
a &= 12.07\text{ ft/sec}^2
\end{aligned}$$

The tension in the wire is

$$\begin{aligned}
T &= \left(0.932\ \frac{\text{lbf-sec}^2}{\text{ft}}\right)a \\
&= \left(0.932\ \frac{\text{lbf-sec}^2}{\text{ft}}\right)\left(12.07\ \frac{\text{ft}}{\text{sec}^2}\right) \\
&= 11.25\text{ lbf} \quad (11\text{ lbf})
\end{aligned}$$

The answer is (A).

50. The effective load is greater than the actual load due to the speed of application (impact). Calculate the effective load from the impact factor equation. Set the height of impact to zero. Since the last term is zero, it is not necessary to actually know the deflection.

$$\begin{aligned}
F_e &= P\left(1 + \sqrt{1 + \frac{2h}{\delta_{\text{st}}}}\right) \\
&= (1000\text{ lbf})\left(1 + \sqrt{1 + \frac{(2)(0\text{ ft})}{\delta_{\text{st}}}}\right) \\
&= 2000\text{ lbf}
\end{aligned}$$

Calculate the weld parameters treating the weld as a line.

$$\begin{aligned}
A_w &= 2L_w = (2)(10\text{ in}) \\
&= 20\text{ in} \\
S_w &= \frac{L_w^2}{3} = \frac{(10\text{ in})^2}{3} \\
&= 33.3\text{ in}^2 \\
M &= (2000\text{ lbf})(2\text{ in}) \\
&= 4000\text{ in-lbf}
\end{aligned}$$

Calculate forces on the weld. The direct shear is

$$\begin{aligned}
V &= P = 2000\text{ lbf} \\
f_s &= \frac{P}{A_w} = \frac{2000\text{ lbf}}{20\text{ in}} \\
&= 100\text{ lbf/in}
\end{aligned}$$

The shear due to bending is

$$\begin{aligned}
f_b &= \frac{M}{S_w} = \frac{4000\text{ in-lbf}}{33.3\text{ in}^2} \\
&= 120\text{ lbf/in}
\end{aligned}$$

Calculate the resolved shear force.

$$\begin{aligned}
f &= \sqrt{f_s^2 + f_b^2} \\
&= \sqrt{\left(100\ \frac{\text{lbf}}{\text{in}}\right)^2 + \left(120\ \frac{\text{lbf}}{\text{in}}\right)^2} \\
&= 156.2\text{ lbf/in}
\end{aligned}$$

Calculate the weld force capacity. The allowable force per inch of weld is 5000 lbf/in. The weld size is $\frac{1}{8}$ in. (The weld throat is smaller than the weld size, but the weld resistance is given per inch of weld size, not weld throat.)

$$f_{\text{allowable}} = \left(5000\ \frac{\text{lbf}}{\text{in}}\right)\left(\frac{\frac{1}{8}\ \text{in}}{1\ \text{in}}\right) = 625\ \text{lbf/in}$$

Calculate the factor of safety.

$$\text{FS} = \frac{f_{\text{allowable}}}{f} = \frac{625\ \frac{\text{lbf}}{\text{in}}}{156.2\ \frac{\text{lbf}}{\text{in}}} = 4.0$$

The answer is (A).

51. Definitions from the American Welding Society are:

Soldering—A joining process wherein coalescence between metal parts is produced by heating to suitable temperatures generally below 800°F and by using nonferrous filler metals having melting temperatures below those of the base metals. The solder is usually distributed between the properly fitted surfaces of the joint by capillary attraction.

Brazing—A group of welding processes wherein coalescence is produced by heating to suitable temperatures above 800°F and by using a nonferrous filler metal having a melting point below those of the base metals. The filler metal is distributed between closely fitted surfaces of the joint by capillary action.

Welding—A localized coalescence of metals wherein coalescence is produced by heating to suitable temperatures, with or without the application of pressure, and with or without the use of filler metal. The filler metal either has a melting point approximately the same as the base metals, or has a melting point below that of the base metals but above 800°F.

Forge Welding—A group of welding processes wherein coalescence is produced by heating in a forge or other furnace and by applying pressure or blows. An example is the hammer welding process previously used in railroad and blacksmith shops.

The answer is (A).

52. Consult tables of fits that are available in most design references. The following table is from ANSI standard B4.1-1967, R1987.

nominal range	class RC 7		
	clearance	standard tolerance limits (10^{-3} in)	
over–to	(10^{-3} in)	hole H9	shaft d8
0.40–0.71	2.0	+1.6	−2.0
	4.6	0	−3.0

The answer is (A).

53. The logarithmic decrement is the natural logarithm of the ratio of two successive amplitudes. It is related to the damping ratio by the equation

$$\ln\frac{x_n}{x_{n+1}} = \frac{2\pi\zeta}{\sqrt{1-\zeta^2}}$$

Squaring both sides and solving for the damping ratio,

$$\left(\ln\frac{x_n}{x_{n+1}}\right)^2 = \frac{4\pi^2\zeta^2}{1-\zeta^2}$$

$$4\pi^2\zeta^2 = \left(\ln\frac{x_n}{x_{n+1}}\right)^2 - \zeta^2\left(\ln\frac{x_n}{x_{n+1}}\right)^2$$

$$\zeta^2\left(4\pi^2 + \left(\ln\frac{x_n}{x_{n+1}}\right)^2\right) = \left(\ln\frac{x_n}{x_{n+1}}\right)^2$$

$$\zeta = \sqrt{\frac{\left(\ln\frac{x_n}{x_{n+1}}\right)^2}{4\pi^2 + \left(\ln\frac{x_n}{x_{n+1}}\right)^2}} = \sqrt{\frac{\left(\ln\frac{0.569\ \text{in}}{0.462\ \text{in}}\right)^2}{4\pi^2 + \left(\ln\frac{0.569\ \text{in}}{0.462\ \text{in}}\right)^2}} = 0.0331\quad(0.033)$$

The answer is (B).

54. Consider the overall geometry of the truss. The tension in the wire is equal to the weight.

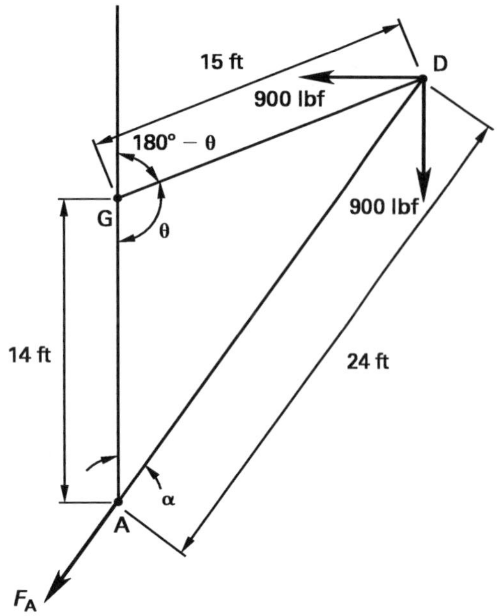

Calculate angles θ and α with the law of cosines.

$$a^2 = b^2 + c^2 - 2bc\cos A$$

$$\cos A = \frac{b^2 + c^2 - a^2}{2bc}$$

$$\begin{aligned}\cos\theta &= \frac{(\text{AG})^2 + (\text{DG})^2 - (\text{AD})^2}{2(\text{AG})(\text{DG})} \\ &= \frac{(14\text{ ft})^2 + (15\text{ ft})^2 - (24\text{ ft})^2}{(2)(14\text{ ft})(15\text{ ft})} \\ &= -0.369\end{aligned}$$

$$\theta = 112°$$

$$\begin{aligned}\cos\alpha &= \frac{(\text{AG})^2 + (\text{AD})^2 - (\text{DG})^2}{2(\text{AG})(\text{AD})} \\ &= \frac{(14\text{ ft})^2 + (24\text{ ft})^2 - (15\text{ ft})^2}{(2)(14\text{ ft})(24\text{ ft})} \\ &= 0.814\end{aligned}$$

$$\alpha = 35.5°$$

For static equilibrium, the sum of the moments about point G is zero. The pin connection of the two-force member AB can support only a force in the direction of the member AB.

$$\begin{aligned}\sum M_G &= 0\text{ ft-lbf} \\ &= (\text{AG})F_A\sin\alpha + (\text{DG})F_w\sin(180° - \theta) \\ &\quad -(\text{DG})F_w\cos(180° - \theta)\end{aligned}$$

$$\begin{aligned}F_A &= \frac{(\text{DG})F_w(\cos(180° - \theta) - \sin(180° - \theta))}{(\text{AG})\sin\alpha} \\ &= \frac{(15\text{ ft})(900\text{ lbf})\left(\begin{array}{c}\cos(180° - 112°) \\ -\sin(180° - 112°)\end{array}\right)}{(14\text{ ft})\sin 35.5°} \\ &= -917.58\text{ lbf}\end{aligned}$$

All members of a truss are two-force members, so connecting joints do not transmit moments, and each force acts in the direction of the member. A force balance around point B gives

$$F_{BG} = F_{BF} = 0\text{ lbf}$$

$$F_A = F_{BC} = -917.58\text{ lbf} \quad (920\text{ lbf, compression})$$

The answer is (C).

55. The equivalent uniform annual cost (EUAC) compares alternatives that have a mix of present, annual, and future costs. The EUAC is the annual cost that is equivalent to the sum of all costs. Present and future costs are converted to equivalent annual costs through the use of cash flow equivalent factors, such as from MERM App. "Cash Flow Equivalent Factors;" these cash flow equivalent factors take interest into account. Then all annual costs and equivalent annual costs are added together to get the EUAC.

The annual electricity cost of the new system is

$$\begin{aligned}C_{\text{elec}} &= \left(\frac{110\text{ hp}}{1.341\ \frac{\text{hp}}{\text{kW-hr}}}\right)\left(1800\ \frac{\text{hr}}{\text{yr}}\right)\left(0.12\ \frac{\$}{\text{kW-hr}}\right) \\ &= \$17{,}720/\text{yr}\end{aligned}$$

The EUAC for the new system is

$$\begin{aligned}\text{EUAC} &= P_{\text{initial}}(A/P,\ 10\%,\ 10) + A_{\text{elec}} + A_{\text{maint}} \\ &\quad -A_{\text{salvage}}(A/F,\ 10\%,\ 10) \\ &= (\$80{,}000)(0.1627) + \$17{,}720 + \$2000 \\ &\quad -(\$10{,}000)(0.0627) \\ &= \$32{,}100 \quad (\$32{,}000)\end{aligned}$$

The answer is (D).

56. The nominal stack height is the sum of the nominal part heights.

$$\begin{aligned} N &= N_1 + N_2 + N_3 \\ &= 0.8\text{ in} + 1.0\text{ in} + 1.2\text{ in} \\ &= 3.0\text{ in} \end{aligned}$$

The stack tolerance is a root sum square combination of parts tolerances.

$$\begin{aligned} T &= \sqrt{T_1^2 + T_2^2 + T_3^2} \\ &= \sqrt{(0.1\text{ in})^2 + (0.2\text{ in})^2 + (0.3\text{ in})^2} \\ &= 0.37\text{ in} \quad (0.4\text{ in}) \end{aligned}$$

The toleranced stack height combines the stack nominal height and tolerance. Given that all part tolerances are normally distributed and $\pm 3\sigma$ from their means, this value represents 99.7% of stack assemblies.

$$N + T = 3.0\text{ in} \pm 0.4\text{ in}$$

The answer is (C).

57. Sum the three-day averages and divide by 8 to get the average weight of a casting.

$$\begin{aligned} \overline{W} &= \frac{\sum W_i}{8} = \frac{\begin{array}{c} 428.0 + 434.7 + 450.7 + 444 \\ +443.3 + 445 + 458.3 + 452 \end{array}}{8} \\ &= 444.5 \end{aligned}$$

Find the square of the difference of each three-day average from this average, and sum the squares.

$$\begin{aligned} \sum (W_i - \overline{W})^2 &= (428 - 444.5)^2 + (434.7 - 444.5)^2 \\ &\quad + (450.7 - 444.5)^2 + (444 - 444.5)^2 \\ &\quad + (443.3 - 444.5)^2 + (445 - 444.5)^2 \\ &\quad + (458.3 - 444.5)^2 + (452 - 444.5)^2 \\ &= 655 \end{aligned}$$

The sample standard deviation is

$$\begin{aligned} s &= \sqrt{\frac{\sum (W_i - W)^2}{n-1}} = \sqrt{\frac{655}{8-1}} \\ &= 9.67 \quad (10) \end{aligned}$$

The answer is (D).

58. The critical path diagram is shown. Each box represents an activity, and the number is the box is the number of days needed to complete the activity. The critical path is the path through the diagram containing the greatest number of days, and its length is the least number of days in which the project can be completed.

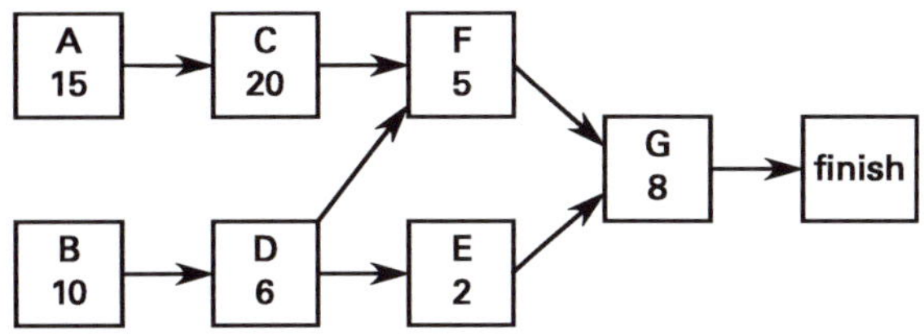

The critical path through this diagram is A-C-F-G-finish. The length of this path is 15 days + 20 days + 5 days + 8 days = 48 days.

The answer is (C).

59. In the plan view (at the top of the illustration), the distance from the centerline to the extent of A is $2\frac{5}{8}$ in. This length is the sum of A, a "valley" of unknown width, and $\frac{7}{8}$ in. The width of the "valley" can be found in the front elevation view (directly below the plan view); it is $\frac{5}{8}$ in. The value of dimension A is $2\frac{5}{8}$ in − $\frac{5}{8}$ in − $\frac{7}{8}$ in = $1\frac{1}{8}$ in.

The answer is (B).

60. Draw a free-body diagram showing only external forces as shown. The force at pin connection A could be in both the x- and y-directions, while the supporting force at roller connection D can only be in the y-direction.

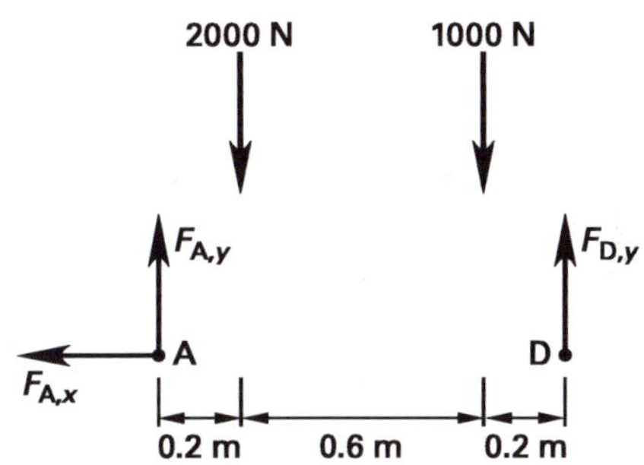

However, the sum of all forces in the x-direction must be zero, so $F_{A,x}$ is zero. Summing forces in the y-direction gives

$$\begin{aligned} \sum F_y &= 0 \\ F_{A,y} + F_{D,y} - 2000\text{ N} - 1000\text{ N} &= 0 \\ F_{A,y} + F_{D,y} &= 3000\text{ N} \end{aligned}$$

Summing the moments around pin A (with counterclockwise moments positive and clockwise moments negative) gives

$$\begin{aligned} \sum M_A &= 0 \\ (1\text{ m})F_{D,y} - (0.2\text{ m})(2000\text{ N}) - (0.8\text{ m})(1000\text{ N}) &= 0 \end{aligned}$$

$$F_{D,y} = \frac{1200 \text{ N·m}}{1 \text{ m}} = 1200 \text{ N}$$

Therefore,

$$\begin{aligned} F_{A,y} &= 3000 \text{ N} - F_{D,y} \\ &= 3000 \text{ N} - 1200 \text{ N} \\ &= 1800 \text{ N} \end{aligned}$$

Draw a free-body diagram of the horizontal element B-E and the forces acting on it.

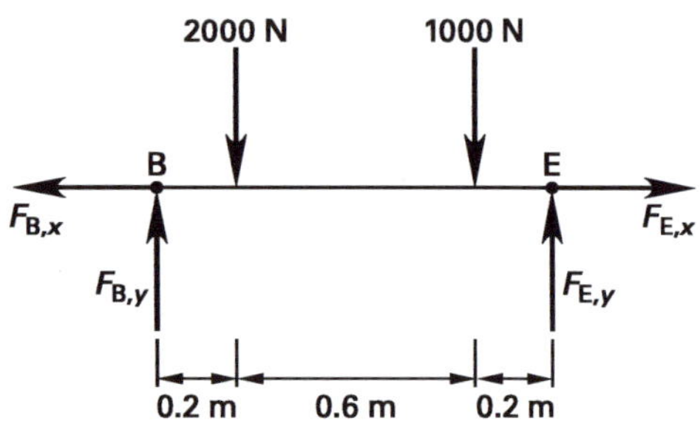

Summing the moments around point B gives

$$\sum M_B = 0$$

$$(1 \text{ m})F_{E,y} - (0.2 \text{ m})(2000 \text{ N}) - (0.8 \text{ m})(1000 \text{ N}) = 0$$

$$F_{E,y} = \frac{1200 \text{ N·m}}{1 \text{ m}} = 1200 \text{ N}$$

Isolate the element A-E with all the forces on it.

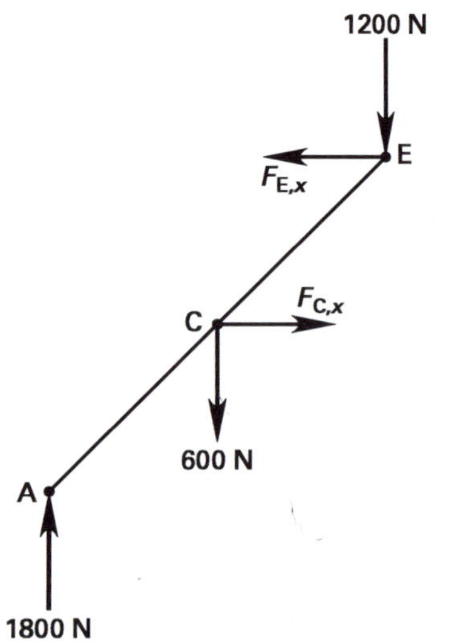

The vertical force on point C is

$$\begin{aligned} F_{C,y} &= F_{A,y} - F_{E,y} = 1800 \text{ N} - 1200 \text{ N} \\ &= 600 \text{ N} \end{aligned}$$

Summing the moments around point C gives

$$\sum M_C = 0$$

$$(0.5 \text{ m})F_{E,x} - (0.5 \text{ m})(1800 \text{ N}) - (0.5 \text{ m})(1200 \text{ N}) = 0$$

$$F_{E,x} = \frac{1500 \text{ N·m}}{0.5 \text{ m}} = 3000 \text{ N}$$

The answer is (D).

61. The projectile motion equation for range with target below is

$$R = \left(\frac{v_0 \cos\phi}{g}\right)\left(v_0 \sin\phi + \sqrt{2gz + v_0^2 \sin^2\phi}\right)$$

In this case, the angle at which the projectile is launched, ϕ, is zero, so this equation reduces to

$$R = \frac{v_0}{g}\sqrt{2gz}$$

R is the horizontal distance and z the vertical distance from the exit point of the horizontal conveyor to the impact point on the lower conveyor. If a vertical line is drawn from the exit point of the upper conveyor to the base of the lower conveyor, and a horizontal line is drawn from the impact point to the vertical line, a right triangle is formed as shown.

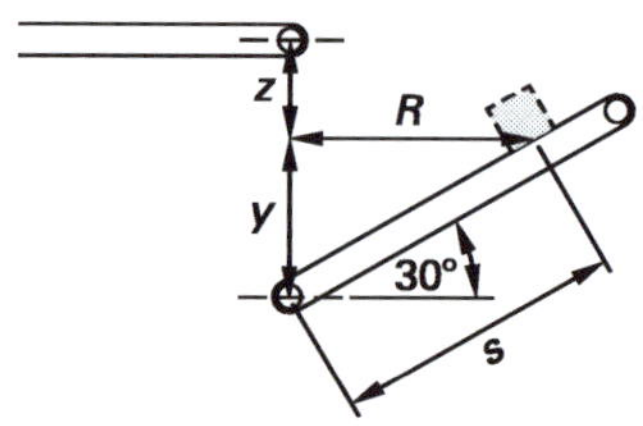

Let y be the side of the triangle opposite the 30° angle, and let s be the hypotenuse. Then,

$$y = s \sin 30° = 0.5s$$

$$z = 10 \text{ m} - y = 10 \text{ m} - 0.5s$$

$$\begin{aligned} s &= \frac{R}{\cos 30°} = \frac{\frac{v_0}{g}\sqrt{2gz}}{0.866} = \frac{v_0}{0.866g}\sqrt{2gz} \\ &= \frac{9 \frac{\text{m}}{\text{s}}}{(0.866)\left(9.81 \frac{\text{m}}{\text{s}^2}\right)}\sqrt{(2)\left(9.81 \frac{\text{m}}{\text{s}^2}\right) \times (10 \text{ m} - 0.5s)} \\ &= (1.059 \text{ s})\sqrt{(2)\left(9.81 \frac{\text{m}}{\text{s}^2}\right)(10 \text{ m} - 0.5s)} \end{aligned}$$

Squaring both sides gives

$$s^2 = (1.121\ \text{s}^2)(2)\left(9.81\ \frac{\text{m}}{\text{s}^2}\right)(10\ \text{m} - 0.5s)$$

$$s^2 + (11.00\ \text{m})s - 219.9\ \text{m}^2 = 0$$

$$s = \frac{-11.00\ \text{m} \pm \sqrt{(11.00\ \text{m})^2 - (4)(1)(-219.9\ \text{m}^2)}}{(2)(1)}$$

$$= 10.32\ \text{m} \quad (10\ \text{m})$$

The answer is (D).

62. Stainless steel is widely used in commercial kitchens, both for its hardness and for its corrosion resistance. Its primary alloying metal is chromium, ideal for hardness, toughness, and rust prevention.

Manganese improves wear resistance. Vanadium improves toughness. Carbon is a basic ingredient in steel, not an alloy for it.

The answer is (C).

63. There are many types of hardness tests. Each has a method of imprinting a shape into the surface of a material with a certain amount of force. In the case of the Brinell and Rockwell tests, the imprint is of a sphere. In the case of the Vickers test, the shape is a square pyramid.

The Charpy test involves a pendulum striker, and the test measures toughness, not hardness.

The answer is (C).

64. Poisson's ratio, ν, is the ratio of lateral strain to axial (or longitudinal) strain.

$$\nu = \frac{e_{\text{lateral}}}{e_{\text{axial}}} = \frac{\dfrac{\Delta D}{D_o}}{\dfrac{\delta}{L_o}}$$

When a material is stretched with an axial strain, the material stretches in size both in the axial direction (the direction of the tension) and in the lateral direction (the cross section).

For almost all types of materials, the Poisson ratio is less than 0.5. For most metals, the Poisson ratio is 0.3. This means that for every unit of axial strain, there is 0.3 of a unit of lateral strain. Hence, for a metal rod under axial tension, the diameter decreases when the length increases.

The answer is (D).

65. The design equation that relates all the parameters of the spring is

$$k = \frac{Gd}{8C^3 n_a}$$

In this equation, G is the shear modulus of the wire, d is the thickness of the wire, and n_a is the number of active coils. C is the spring index, which is equal to D/d, where D is the diameter of the spring. The needed spring rate, k, can be found from the deflection equation.

$$k = \frac{P}{\delta} = \frac{5000\ \text{lbf}}{6\ \text{in}} = 833.3\ \text{lbf/in}$$

Use the design equation and solve for the spring index.

$$C = \sqrt[3]{\frac{Gd}{8kn_a}} = \sqrt[3]{\frac{\left(12.0 \times 10^6\ \frac{\text{lbf}}{\text{in}^2}\right)(1\ \text{in})}{(8)\left(833.3\ \frac{\text{lbf}}{\text{in}}\right)(10)}}$$

$$= 5.64$$

The needed diameter is

$$D = Cd = (5.64)(1\ \text{in})$$

$$= 5.64\ \text{in} \quad (5.6\ \text{in})$$

The answer is (C).

66. The equivalent uniform annual cost (EUAC) is a convenient way to compare alternatives that have a mix of differing costs. The EUAC is the annual cost that is equivalent to the sum of all costs. Present and future costs are converted to equivalent annual costs through the use of cash flow equivalent factors, such as from MERM App. "Cash Flow Equivalent Factors;" these cash flow equivalent factors take interest into account. Then all annual costs and equivalent annual costs are added together to get the EUAC.

Letting H be the number of fully loaded hours per year, the annual operating cost is

$$A_{\text{oper}} = \left(\frac{P}{\eta}\right) C_{\text{elec}} H$$

$$A_{\text{oper,A}} = \left(\frac{\dfrac{100\ \text{hp}}{1.341\ \dfrac{\text{hp}}{\text{kW-hr}}}}{0.85}\right)\left(0.11\ \frac{\$}{\text{kW-hr}}\right)H = (\$9.65)H$$

$$A_{\text{oper,B}} = \left(\frac{\dfrac{100\ \text{hp}}{1.341\ \dfrac{\text{hp}}{\text{kW-hr}}}}{0.92}\right)\left(0.11\ \frac{\$}{\text{kW-hr}}\right)H = (\$8.92)H$$

The EUAC for each model is

$$\begin{aligned} \text{EUAC} &= P_{\text{initial}}(A/P,\ 8\%,\ 10) + A_{\text{maint}} + A_{\text{oper}} \\ \text{EUAC}_{\text{A}} &= (\$2500)(0.1490) + \$100 + (\$9.65)H \\ &= \$472.50 + (\$9.65)H \\ \text{EUAC}_{\text{B}} &= (\$3300)(0.1490) + \$60 + (\$8.92)H \\ &= \$551.70 + (\$8.92)H \end{aligned}$$

To find the value of H that will make both EUACs equal, set the two equations equal to each other and solve for H.

$$\begin{aligned} \$472.50 + (\$9.65)H &= \$551.70 + (\$8.92)H \\ (\$9.65)H - (\$8.92)H &= \$551.70 - \$472.50 \\ H &= \frac{\$551.70 - \$472.50}{\$9.65 - \$8.92} \\ &= 108.5 \quad (110) \end{aligned}$$

The answer is (A).

67. From MERM App. "Viscosities Versus Temperature of Hydrocarbons, Lubricants, Fuels, and Heat Transfer Fluids," the kinematic viscosity of SAE 10 oil at 150°F is 15 cSt. Convert centistokes to reyns by multiplying by 1.45×10^{-7} and the specific gravity.

$$\begin{aligned} \mu_{\text{reyns}} &= (15\ \text{cSt})\left(1.45 \times 10^{-7}\ \frac{\text{reyns}}{\text{cSt}}\right)(0.85) \\ &= 1.85 \times 10^{-6}\ \text{reyns} \end{aligned}$$

The diametrical clearance is twice the radial clearance.

$$\begin{aligned} c_d &= 2c_r = (2)(0.004\ \text{in}) \\ &= 0.008\ \text{in} \end{aligned}$$

The rotational speed in revolutions per second is

$$\begin{aligned} n_{\text{rps}} &= \frac{1000\ \dfrac{\text{rev}}{\text{min}}}{60\ \dfrac{\text{sec}}{\text{min}}} \\ &= 16.67\ \text{rev/sec} \end{aligned}$$

Use the Petroff equation to find the frictional torque.

$$\begin{aligned} T_{f,\text{in-lbf}} &= \frac{\pi^2 \mu_{\text{reyns}} d^3 L n_{\text{rps}}}{c_d} \\ &= \frac{\pi^2(1.85 \times 10^{-6}\ \text{reyns})(6\ \text{in})^3 \times (9\ \text{in})\left(16.67\ \dfrac{\text{rev}}{\text{sec}}\right)}{0.008\ \text{in}} \\ &= 73.88\ \text{in-lbf} \quad (74\ \text{in-lbf}) \end{aligned}$$

The answer is (C).

68. The mechanical energy of the flywheel is fully dissipated into heat during the braking action, so the heat dissipated during the braking is equal to the mechanical energy of the flywheel.

The angular velocity of the flywheel is

$$\begin{aligned} \omega &= \left(2\pi\ \frac{\text{rad}}{\text{rev}}\right)\left(\frac{500\ \dfrac{\text{rev}}{\text{min}}}{60\ \dfrac{\text{sec}}{\text{min}}}\right) \\ &= 52.4\ \text{rad/sec} \end{aligned}$$

The kinetic energy stored in the flywheel is

$$\begin{aligned} E_{k,r} &= \frac{I\omega^2}{2g_c} = \frac{(mk^2)\omega^2}{2g_c} \\ &= \frac{(450\ \text{lbm})\left(\dfrac{6\ \text{in}}{12\ \dfrac{\text{in}}{\text{ft}}}\right)^2\left(52.4\ \dfrac{\text{rad}}{\text{sec}}\right)^2}{(2)\left(32.2\ \dfrac{\text{lbm-ft}}{\text{lbf-sec}^2}\right)} \\ &= 4796.6\ \text{ft-lbf} \end{aligned}$$

Divide the kinetic energy by Joule's constant to convert to thermal energy.

$$Q = \frac{E_{k,r}}{J} = \frac{4796.6 \text{ ft-lbf}}{778.17 \ \frac{\text{ft-lbf}}{\text{Btu}}} = 6.164 \text{ Btu} \quad (6 \text{ Btu})$$

The answer is (A).

69. From MERM Table "Approximate Values of Poisson's Ratio," the value of Poisson's ratio, ν, for cast iron is commonly taken as 0.27. From the MERM App. "Thermal Properties of Selected Solids," the density of cast iron is 455 lbm/ft^3. The outer radius of the flywheel, r_o, is 4 in.

The angular velocity of the flywheel is

$$\omega = \left(2\pi \ \frac{\text{rad}}{\text{rev}}\right)\left(\frac{20{,}000 \ \frac{\text{rev}}{\text{min}}}{60 \ \frac{\text{sec}}{\text{min}}}\right) = 2094 \text{ rad/sec}$$

In a solid, rotating disk, the radial and tangential stresses at a radius r are

$$\sigma_r = \left(\frac{3+\nu}{8}\right)\left(\frac{\rho\omega^2}{g_c}\right)(r_o^2 - r^2)$$

$$\sigma_t = \left(\frac{\rho\omega^2}{g_c}\right)\left(\left(\frac{3+\nu}{8}\right)r_o^2 - \left(\frac{1+3\nu}{8}\right)r^2\right)$$

However, tangential and radial stresses are equal and greatest at the center—that is, when r is zero. Therefore,

$$\sigma_r = \sigma_t = \left(\frac{3+\nu}{8}\right)\left(\frac{\rho\omega^2}{g_c}\right)r_o^2$$

$$= \left(\frac{3+0.27}{8}\right)\left(\frac{\left(455 \ \frac{\text{lbm}}{\text{ft}^3}\right)\left(2094 \ \frac{\text{rad}}{\text{sec}}\right)^2}{32.2 \ \frac{\text{lbm-ft}}{\text{lbf-sec}^2}}\right)\left(\frac{4 \text{ in}}{12 \ \frac{\text{in}}{\text{ft}}}\right)^2$$

$$= 2{,}814{,}000 \text{ lbf/in}^2 \quad (2{,}800{,}000 \text{ psi})$$

The answer is (C).

70. The pitch of a thread is the number of inches per thread. The thread nomenclature lists the size of the bolt shaft followed by the number of threads per inch. For this bolt, there are 20 threads per inch, so the number of inches per thread is 1/20, or 0.05 in/thread.

The answer is (B).

71. The minimum wall thickness for a cylindrical shell is

$$t = \frac{pR}{SE - 0.6p} = \frac{\left(1000 \ \frac{\text{lbf}}{\text{in}^2}\right)(19 \text{ in})}{\left(20{,}000 \ \frac{\text{lbf}}{\text{in}^2}\right)(0.85) - (0.6)\left(1000 \ \frac{\text{lbf}}{\text{in}^2}\right)} = 1.159 \text{ in} \quad (1.2 \text{ in})$$

The answer is (B).

72. The areas of the rivet holes are

$$A_{8\text{mm}} = \frac{\pi d^2}{4} = \frac{\pi(8 \text{ mm})^2}{4} = 50.2 \text{ mm}^2$$

$$A_{10\text{mm}} = \frac{\pi d^2}{4} = \frac{\pi(10 \text{ mm})^2}{4} = 78.5 \text{ mm}^2$$

The distance from the geometric center to the center of each hole is

$$r = \frac{\sqrt{(30 \text{ mm})^2 + (40 \text{ mm})^2}}{2} = 25 \text{ mm}$$

Because r is the same for each hole, the polar moment of inertia is

$$J = \sum r_i^2 A_i = r^2 \sum A_i$$

$$= (25 \text{ mm})^2\left(50.2 \text{ mm}^2 + 78.5 \text{ mm}^2 + 50.2 \text{ mm}^2 + 78.5 \text{ mm}^2\right)$$

$$= 160{,}875 \text{ mm}^4 \quad (160{,}000 \text{ mm}^4)$$

The answer is (D).

73. The preload on the bolt is tension to the bolt, but compression to the clamping force. With the addition of a 200 lbf tension load, the clamping force is reduced by 200 lbf to

$$F = 500 \text{ lbf} - 200 \text{ lbf} = 300 \text{ lbf}$$

The tension load on the bolt changes only slightly.

The answer is (B).

74. The applied load, F, is assumed to be carried in shear by the effective weld throat. The effective throat size, t_e, is related to the weld size, y.

$$\begin{aligned} t_e &= \frac{\sqrt{2}}{2}y = \left(\frac{\sqrt{2}}{2}\right)(0.25\ \text{in}) \\ &= 0.1768\ \text{in} \end{aligned}$$

The total length of the welds is

$$\begin{aligned} b &= (2)(2\ \text{in}) \\ &= 4\ \text{in} \end{aligned}$$

The shear stress is

$$\begin{aligned} \tau &= \frac{F}{bt_e} = \frac{10{,}000\ \text{lbf}}{(4\ \text{in})(0.1768\ \text{in})} \\ &= 14{,}140\ \text{lbf/in}^2 \end{aligned}$$

The factor of safety is

$$\begin{aligned} \text{FS} &= \frac{S_y}{\tau} = \frac{30{,}000\ \frac{\text{lbf}}{\text{in}^2}}{14{,}140\ \frac{\text{lbf}}{\text{in}^2}} \\ &= 2.122 \quad (2.1) \end{aligned}$$

The answer is (D).

75. The amount that the diameter is to be reduced is 2 in − 1.25 in = 0.75 in. The depth of cut is 0.125 in, so the number of passes needed is 0.75 in/0.125 in = 6. The speed of the lathe is based on the first pass.

$$\begin{aligned} n &= \frac{\text{v}}{\pi d} = \frac{\left(100\ \frac{\text{ft}}{\text{min}}\right)\left(12\ \frac{\text{in}}{\text{ft}}\right)}{\pi(2\ \text{in})} \\ &= 191\ \text{rev/min} \end{aligned}$$

The feed rate in inches per minute is

$$\begin{aligned} f &= \left(0.02\ \frac{\text{in}}{\text{rev}}\right)\left(191\ \frac{\text{rev}}{\text{min}}\right) \\ &= 3.8\ \text{in/min} \end{aligned}$$

The time for one pass is

$$\begin{aligned} t &= \frac{L}{f} = \frac{12\ \text{in}}{3.8\ \frac{\text{in}}{\text{min}}} \\ &= 3.1\ \text{min} \end{aligned}$$

The speed of the lathe does not change, so the feed rate and the time for one pass also do not change. The time for six passes is

$$\begin{aligned} t_{\text{total}} &= 6t = (6)(3.1\ \text{min}) \\ &= 18.6\ \text{min} \quad (19\ \text{min}) \end{aligned}$$

The answer is (D).

76. The rotational speed is

$$\begin{aligned} n &= \frac{\text{v}}{\pi d} = \frac{\left(70\ \frac{\text{ft}}{\text{min}}\right)\left(12\ \frac{\text{in}}{\text{ft}}\right)}{\pi(2\ \text{in})} \\ &= 133.8\ \text{rev/min} \end{aligned}$$

The feed rate is

$$\begin{aligned} f &= (3\ \text{teeth})\left(0.012\ \frac{\text{in}}{\text{tooth}}\right)\left(133.8\ \frac{\text{rev}}{\text{min}}\right) \\ &= 4.817\ \text{in/min} \quad (4.8\ \text{in/min}) \end{aligned}$$

The answer is (A).

77. The Society of Automobile Engineers and the American Iron and Steel Institute (IV) have devised similar methods of identifying steels by their carbon content and type and percentage of alloys.

The American Steel Construction Institute (III) wrote *Specification for Structural Joints*, the guiding standard for sizing bolted connections.

The *ASME Boiler and Pressure Vessel Code* is an American Society of Mechanical Engineers (II) standard that regulates the design and construction of boilers and pressure vessels.

The International Organization for Standardization and the American National Standard Institute (I) have combined to identify the engineering tolerances of inner and outer features of journal bearings, linear bearings, thrust bearings, bushings, ball bearings, roller bearings, housings, cylinder bores, drilled holes, linear and precision shafts, and pistons according to a system of "fits" that includes clearance, transition, and interference fits.

The answer is (D).

78. The Society of the Plastics Industry introduced the Resin Identification Code (RIC) system in 1988, primarily to identify plastics for recycling.

There are seven code numbers:

1. polyethylene terephthalate (PET)
2. high-density polyethylene (HDPE)
3. polyvinyl chloride (PVC)
4. low-density polyethylene (LDPE)
5. polypropylene (PP)
6. polystyrene (PS)
7. all others

The answer is (A).

79. Polystyrene (PS) should be marked with the symbol

Acrylonitrile butadiene styrene (ABS), polycarbonate (PC), and nylon should all be marked with the symbol

The answer is (B).

80. Calculate the mean of the sample data.

$$\begin{aligned}\overline{x} &= \frac{0.010 + 0.014 + 0.002 + (-0.020) + (-0.001)}{5} \\ &= 0.001\end{aligned}$$

Calculate the sample standard deviation.

$$\begin{aligned}s &= \sqrt{\frac{\sum(x_i - \overline{x})^2}{n-1}} \\ &= \sqrt{\frac{\begin{array}{l}(0.010 - 0.001)^2 + (0.014 - 0.001)^2 \\ \quad +(0.002 - 0.001)^2 + (-0.020 - 0.001)^2 \\ \quad +(-0.001 - 0.001)^2\end{array}}{5-1}} \\ &= 0.013\end{aligned}$$

Since the sample standard deviation is 0.013 and the allowable deviation is 0.03, the allowable deviation represents $0.03/0.013 = 2.3$ times the sample standard deviation. From MERM App. "Areas Under the Standard Normal Curve," for $z = 2.3$, the one-sided value for the percentage that falls between zero and z is 0.4893. In this case, the percentage wanted is not just that between zero and z, but the two-sided value between $-z$ and z. This percentage is twice 0.4893, or 0.9786.

The home position will differ from zero by 0.03 or less 97.86% of the time. Therefore, the percentage of the time that the home position will differ from zero by more than 0.03 is

$$100\% - 97.86\% = 2.14\% \quad (2.1\%)$$

The answer is (C).